Dmytro Kresan

Fluctuations of particle yield ratios

Dmytro Kresan

Fluctuations of particle yield ratios

Event-by-event fluctuations of particle yield ratios in heavy-ion collisions at 20 - 158 AGeV

Südwestdeutscher Verlag für Hochschulschriften

Impressum/Imprint (nur für Deutschland/ only for Germany)
Bibliografische Information der Deutschen Nationalbibliothek: Die Deutsche Nationalbibliothek verzeichnet diese Publikation in der Deutschen Nationalbibliografie; detaillierte bibliografische Daten sind im Internet über http://dnb.d-nb.de abrufbar.

Alle in diesem Buch genannten Marken und Produktnamen unterliegen warenzeichen-, marken- oder patentrechtlichem Schutz bzw. sind Warenzeichen oder eingetragene Warenzeichen der jeweiligen Inhaber. Die Wiedergabe von Marken, Produktnamen, Gebrauchsnamen, Handelsnamen, Warenbezeichnungen u.s.w. in diesem Werk berechtigt auch ohne besondere Kennzeichnung nicht zu der Annahme, dass solche Namen im Sinne der Warenzeichen- und Markenschutzgesetzgebung als frei zu betrachten wären und daher von jedermann benutzt werden dürften.

Verlag: Südwestdeutscher Verlag für Hochschulschriften Aktiengesellschaft & Co. KG
Dudweiler Landstr. 99, 66123 Saarbrücken, Deutschland
Telefon +49 681 37 20 271-1, Telefax +49 681 37 20 271-0
Email: info@svh-verlag.de
Zugl.: Frankfurt, Goethe Uni, Diss., 2010

Herstellung in Deutschland:
Schaltungsdienst Lange o.H.G., Berlin
Books on Demand GmbH, Norderstedt
Reha GmbH, Saarbrücken
Amazon Distribution GmbH, Leipzig
ISBN: 978-3-8381-1714-0

Imprint (only for USA, GB)
Bibliographic information published by the Deutsche Nationalbibliothek: The Deutsche Nationalbibliothek lists this publication in the Deutsche Nationalbibliografie; detailed bibliographic data are available in the Internet at http://dnb.d-nb.de.

Any brand names and product names mentioned in this book are subject to trademark, brand or patent protection and are trademarks or registered trademarks of their respective holders. The use of brand names, product names, common names, trade names, product descriptions etc. even without a particular marking in this works is in no way to be construed to mean that such names may be regarded as unrestricted in respect of trademark and brand protection legislation and could thus be used by anyone.

Publisher: Südwestdeutscher Verlag für Hochschulschriften Aktiengesellschaft & Co. KG
Dudweiler Landstr. 99, 66123 Saarbrücken, Germany
Phone +49 681 37 20 271-1, Fax +49 681 37 20 271-0
Email: info@svh-verlag.de

Printed in the U.S.A.
Printed in the U.K. by (see last page)
ISBN: 978-3-8381-1714-0

Copyright © 2010 by the author and Südwestdeutscher Verlag für Hochschulschriften Aktiengesellschaft & Co. KG and licensors
All rights reserved. Saarbrücken 2010

Contents

1 Introduction   5
- 1.1 Low-mass vector mesons 7
- 1.2 Charm production 8
- 1.3 Strangeness in dense matter 9
- 1.4 Event-by-event fluctuations 9
- 1.5 The UrQMD model 12

2 The NA49 experiment   15
- 2.1 The magnets 16
- 2.2 The TPC tracking system 16
  - 2.2.1 Principles of operation 16
  - 2.2.2 Readout chambers 17
- 2.3 Veto calorimeter 18
  - 2.3.1 Centrality determination 19

3 Data analysis   21
- 3.1 Inclusive fit 22
- 3.2 Event-by-event fit 35
- 3.3 Extraction of dynamical fluctuations 37
- 3.4 Event and track cuts 39
- 3.5 Systematic errors due to track selection 39

4 Results of the data analysis   41
- 4.1 Energy dependence 41
- 4.2 Dependence on centrality bin size 48
  - 4.2.1 Central events 49
  - 4.2.2 Semi-peripheral events 53

		4.3	Centrality dependence	57
5	**Discussion of the results**			**63**
	5.1	UrQMD simulations		71
		5.1.1	Influence of detector acceptance	71
		5.1.2	dE/dx resolution	73
		5.1.3	Centrality dependence	74
		5.1.4	Resonance contributions in UrQMD	78
	5.2	Simulations with phase space models		80
		5.2.1	Influence of the $K^*(892)$ decay	81
		5.2.2	Influence of the $\phi(1020)$ decay	81
	5.3	Scaling of the dynamical fluctuations		84
	5.4	Conclusion		86
6	**The CBM experiment**			**89**
	6.1	Detector design		90
		6.1.1	Micro vertex detector (MVD)	90
		6.1.2	Silicon tracking system (STS)	91
		6.1.3	Ring imaging cherenkov (RICH) detectors	92
		6.1.4	Transition radiation detector(TRD)	92
		6.1.5	Resistive plate chambers (RPC)	93
		6.1.6	Electromagnetic calorimeter (ECAL)	93
		6.1.7	Muon chambers (MUCH)	94
		6.1.8	Projectile spectator detector (PSD)	94
	6.2	Event reconstruction		94
		6.2.1	STS tracking	95
		6.2.2	TRD tracking	96
		6.2.3	Global tracking	97
	6.3	Hadron identification		97
		6.3.1	Hadron ID with CBM at SIS 100	104
	6.4	Event-by-event fluctuations of the kaon to pion yield ratio		106
		6.4.1	Simulations with UrQMD	106
		6.4.2	Sensitivity test with Toy Model	111
	6.5	Conclusion		111

A Dependence on centrality bin size		**113**
A.1	Central events	113
A.2	Semi-peripheral events	117
B Centrality dependence		**123**
B.1	5% bin width	123
B.2	10% bin width	137
Bibliography		**147**

Chapter 1

Introduction

The exploration of the phase diagram of strongly interacting matter is one of the most challenging fields of modern high-energy physics [1, 2]. The current status of our knowledge is summarised in figure 1.1.

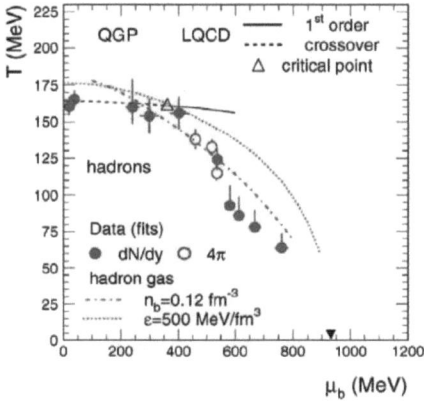

Figure 1.1: Current status of the experimental scan of the QCD phase diagram [2].

The circles show the freeze-out points of different collision energies. The solid line shows the first order phase transition and the dashed line the crossover region as predicted by [3]. The line of the first order phase transition ends with the critical point, which is shown as triangle. The location of this critical point is obtained from lattice calculations and is currently expected around $T=160$ MeV and $\mu_B=400$ MeV [3]. Of particular interest is the transition from hadronic to partonic degrees of freedom which is expected to occur at high temperatures and/or high baryon densities. Both phases played an important role in the early universe and possibly exist in the core of neutron stars [4]. The discovery of this phase transition would shed light on two fundamental but still puzzling aspects of Quantum Chromo Dynamics (QCD): confinement and chiral symmetry breaking. In particular, at high baryon densities one expects new phases of strongly interacting matter [5]. The scientific progress in this exciting field, QCD at high baryon densities, is driven by new experimental data.

In order to study the dynamics of strongly interacting matter far from its ground state, laboratory experiments are performed with high-energy nucleus-nucleus collisions. By varying the energy of heavy-ion collisions, one probes different regions of the QCD phase diagram. For the intermediate energies (AGS, SIS18, lower SPS energies) the system freezes out at moderate temperatures and high baryon densities, while at higher energies (top SPS energies, RHIC, future LHC) very high temperatures and low baryon densities are reached. The conditions inside the transiently existing fireball, created in A + A collisions are reflected in the abundances and phase-space distributions of the emitted hadrons. Important information on the early phase of the collision is provided by the quark flavor of the observed hadrons. In particular, hadrons containing strange or charmed quarks are regarded as sensitive diagnostic probes of the collision dynamics. The hadrons are either observed directly or identified via their hadronic or leptonic decay products.

Lattice QCD calculations at vanishing baryochemical potential and finite temperature predict the formation of a quark-gluon plasma above energy densities of about 1 GeV/fm^3 [6]. Such conditions may be created in heavy collisions between heavy nuclei already at bombarding energies of about 10 AGeV [7, 8]. Recent lattice QCD calculations at finite baryon chemical potential predict a critical endpoint of the deconfinement phase transition at μ_B ≈400 MeV and T≈160 MeV [9, 10]. The major challenge is to find diagnostic probes which are connected to the chiral symmetry restoration and to the deconfinement phase transition. A signature for the onset of chiral symmetry restoration might be the observation of in-medium modifications of hadrons. In-medium effects have been found for pions in deeply bound pionic atoms [11] and for kaons and antikaons produced in heavy-ion collisions at SIS energies [12, 13, 14]. Very promising observables are also short-lived vector mesons decaying into dilepton pairs inside the fireball. An enhanced yield of low-mass dilepton pairs has been found in central heavy-ion collisions [15, 16]. This observation has triggered an enormous theoretical activity on in-medium modifications of ρ-mesons and the relation to chiral symmetry restoration. A further promising candidate for a probe of in-medium modifications is open charm produced at very high baryon densities [17].

Enormous experimental and theoretical efforts have been and are still devoted to the search and investigation of the deconfined phase of strongly interacting matter, the quark-gluon plasma (QGP). The discovery of the QGP was announced by CERN in 2000 [18]. The arguments given in the press release were essentially based on experimental findings such as enhanced production of low-mass dilepton pairs, J/ψ suppression and the fact that an analysis of particle multiplicities yields a chemical freeze-out temperature of about 170 MeV which is very close to the critical temperature. Several years later, new and complementary experimental results obtained at RHIC were interpreted as signatures of the QGP [19]. The main observations of the RHIC experiments include large values of the elliptic flow in agreement with hydrodynamical calculations, the suppression of high p_t particles and quark number scaling of flow observables.

Many signals have been proposed and are still under discussion, although the hope for finding a "smoking gun" has not yet become true. The discovery of the critical endpoint or signals of a first order phase transition would be such a direct indication for the existence of a new phase. Theoretical investigations suggest that particle density fluctuations occur in the vicinity of the critical endpoint, which might be observed experimentally as nonstatistical event-by-event fluctuations of observables [20]. This phenomenon was also seen in lattice QCD calculations [10]. The fluctuation signal should show up around a certain beam energy.

In this section we will briefly review the existing data which are relevant for the study of the high baryon density region of the QCD phase diagram which is the program for the future CBM experiment at FAIR.

1.1 Low-mass vector mesons

The in-medium spectral function of short-lived vector mesons can be measured via their decay into dilepton pairs. Since leptons are essentially unaffected by the passage through the high-density matter, they provide, as a penetrating probe, almost undistorted information on the conditions in the interior of the collision zone [21].

However, dilepton pairs from vector meson decays are difficult to measure due to the very small branching ratios and the large combinatorial background in heavy-ion collisions. The challenge is to identify the electrons or muons unambiguously, measure their momentum better than $\Delta p/p = 2\%$ and to efficiently suppress the background which stems mainly from Dalitz decays and gamma conversion. The test of theoretical predictions requires high statistics data measured with large acceptance spectrometers.

1.2 Charm production

Particles containing heavy quarks like charm are produced in the early stage of the collision. At FAIR, open and hidden charm production will be studied at beam energies close to the kinematical threshold, and the production mechanisms of D and J/ψ mesons will be sensitive to the conditions inside the early fireball. The anomalous suppression of charmonium due to screening effects in the Quark Gluon Plasma (QGP) was predicted to be an experimental signal of the QGP [22]. J/ψ mesons were measured via their decay into muon pairs in heavy-ion collisions at top CERN-SPS energies by the NA50 experiment. It was found that the J/ψ yield decreases relative to the Drell-Yan yield with increasing collision centrality [23], the so called J/ψ suppression, which was considered as a signal for the QGP discovery. The interpretation of the NA50 results is subject of an ongoing debate as the data can be explained also within hadronic scenarios. Data on open charm will be useful for pinning down the source of the J/ψ suppression, and provide an independent probe of the QGP. The next generation experiments aim at the measurement of open charm which will provide further constraints.

Moreover, the effective masses of D-mesons - a bound state of heavy charm quark and a light quark - are expected to be modified in dense matter similarly to those of kaons. Such a change would be reflected in the relative abundance of charmonium ($c\bar{c}$) and D-mesons [17]. D-mesons can be identified via their decay into kaons and pions ($D^0 \rightarrow \pi K$, $D^{\pm} \rightarrow \pi\pi K$). The experimental challenge is to measure the displaced vertex of kaon-pion pairs with an accuracy of better than 100 μm in order to suppress the large combinatorial background caused by promptly emitted kaons and pions.

1.3 Strangeness in dense matter

One of the early predictions for a QGP signal was an increased production of strangeness in the deconfined phase resulting in an enhanced yield of strange particles after hadronization [24]. This effect was expected to be even more pronounced for multistrange hyperons. Indeed, the NA57 experiment observed that the multiplicity of Ξ and Ω hyperons per participant is higher in nucleus-nucleus collisions than in proton-proton (or proton-Beryllium) collisions [25]. Moreover, the enhancement increases according to the s-hierarchy. Again, the interpretation of these results is still under discussion. An intriguing finding is that the slope of the Ω kinetic energy distribution is steeper than expected. This indicates that the Ω hyperon - which consists only of strange quarks - might freeze out very early. Recently, data on the excitation function of strangeness production measured by NA49 have revived the discussion on the role of strangeness as a signature for a deconfinement phase transition [26]. Of particular interest is the peak (so called "horn") in the dependence of the K^+/π^+ ratio on the projectile energy, which appears at $20A$-$30A$ GeV and is not reproduced within statistical and transport models. This peak coincides with the increase of the event-by-event kaon to pion ratio fluctuations at lower SPS nergies. Whether these observations are connected to the critical point of the QCD phase diagram requires further studies.

1.4 Event-by-event fluctuations

In the vicinity of the deconfinement phase transition, critical density fluctuations have been predicted to cause non-statistical event-by-event fluctuations of experimental observables. In particular, the study of event-by-event fluctuations in the hadro-chemical composition of the particle source offers the possibility to directly observe effects of a phase transition. Depending on the nature and the order of the phase transition one expects anomalies in the energy dependence of event-by-event fluctuations [20]. Ideally, a sudden non-monotonous change in the dynamical fluctuations measured as a function of beam energy would be a signal of the critical endpoint.

From lattice QCD calculations one can obtain the dependence of the quark number susceptibility on T and μ_q [27], which is shown in figure 1.2.

At the critical temperature a peak in the quark number susceptibility develops with increasing μ_q, i.e. approaching the critical point. As susceptibilities are connected to

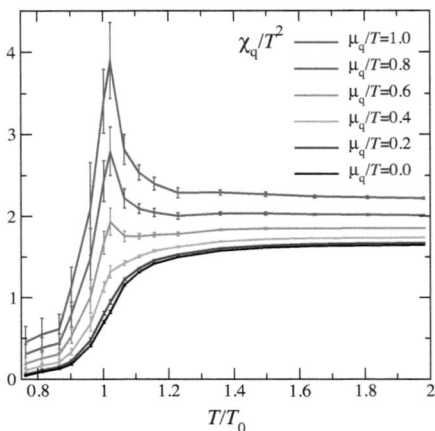

Figure 1.2: The baryon number susceptibility as a function of T and for different μ_B [27].

fluctuations, a non-monotonic energy dependence of the fluctuation signal in the vicinity of the critical point is expected.

The NA49 experiment did measurements of the energy and centrality dependencies of multiplicity fluctuations in A + A collisions in the SPS energy range [28, 29] and mean transverse momentum fluctuations as a function of energy and centrality [30, 31]. From the comparison of the dependencies of these observables to an effect connected to the critical point it can be concluded that the measured signal is far to low than the expected one. Which means that no critical phenomena is observed so far. Maximum of fluctuations is observed in peripheral events, which indicates an important role of geometrical effects.

One can look at strangeness fluctuations by measuring for example kaons. Particle ratios are more preferable since volume fluctuations are canceled in the ratio to first order. Thus one conciders dynamical fluctuations of such observables as the event-by-event kaon to pion, proton to pion and kaon to proton ratios. Event-by-event fluctuations of particle yield ratios have been studied by the NA49 (CERN SPS energy range) and STAR (RHIC energy range) experiments. The dependence of the dynamical fluctuations of the kaon to pion ratio on the center of mass energy of central heavy ion collisions, published by NA49 [32] and STAR [33], is shown in figure 1.3.

An increase of the dynamical fluctuations of the kaon to pion ratio towards lower energies is observed and requires further understanding and interpretation.

1.4. EVENT-BY-EVENT FLUCTUATIONS

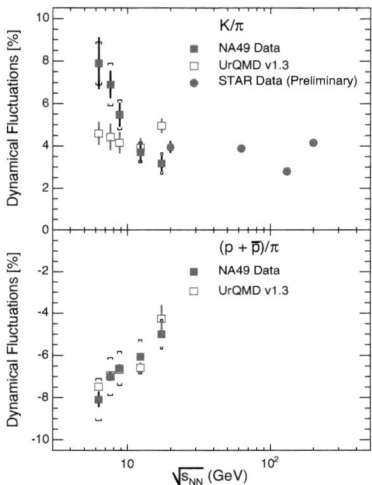

Figure 1.3: Dynamical fluctuations of the kaon to pion and proton to pion ratios as a function of the center of mass energy of central heavy ion collisions [32, 33].

The centrality dependence of the fluctuation signal has been analysed by the STAR collaboration [34]. The observed dependence is shown in figure 1.4.

A scaling with the dn/dy yields at midrapidity was observed. However, the NA49 energy dependent data do not follow this trend. An open issue is here, that the phase space acceptance of NA49 changes with energy.

In order to better understand the available data on the particle ratio fluctuations a measurement of the centrality dependence of dynamical fluctuations of NA49 at one beam energy will be needed. They allow to investigate the multiplicity dependence without changing the acceptance at the same time. This should answer such question as for the multiplicity dependence of this observable.

In this work central and minimum bias Pb + Pb collisions in the CERN SPS beam energy range were analysed with respect to the event-by-event fluctuations of particle yield ratios. In chapter 2 a brief introduction to the NA49 experiment will be presented with focus on the detector components used in this analysis. In chapter 3 the details of the

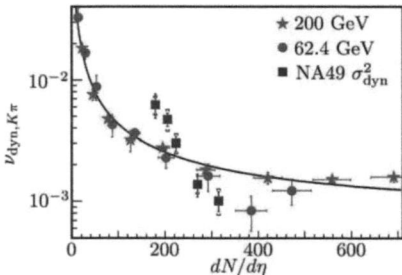

Figure 1.4: Dynamical fluctuations of the kaon to pion ratio as a function of the centrality of heavy ion collisions, published by the STAR collaboration [34].

NA49 data analysis will be discussed. The results of the data analysis will be presented in chapter 4 with a particular focus on the centrality dependence of the particle ratio fluctuations in Pb + Pb collisions at 158A GeV beam energy. In chapter 5 results will be discussed and compared to models. The proposed scaling of dynamical fluctuations with average particle multiplicities will be studied. Description and results of a feasibility study of the measurement of particle ratio fluctuations with the future CBM experiment will be presented in chapter 6.

1.5 The UrQMD model

Currently there are no microscopic transport models with an implemented phase transition and critical point. Thus theoretical predictions for the signal of the critical point can be estimated only from lattice calculations. This makes any interpretation of the measured signal, as the increase of the kaon to pion ratio fluctuations at lower SPS energies [32] and centrality dependence at RHIC energies [34], difficult.
The Ultrarelativistic Quantum Molecular Dynamics (UrQMD) model is a microscopic model used to simulate (ultra)relativistic heavy ion collisons in the energy range from Bevalac and SIS up to AGS, SPS and RHIC [35, 36, 37]. Main goals are to gain understanding about the following physical phenomena within a single transport model:

- Creation of dense hadronic matter at high temperatures
- Properties of nuclear matter, Δ and Resonance matter

1.5. THE URQMD MODEL

- Creation of mesonic matter and of anti-matter
- Creation, modification and destruction of strangeness in matter

Projectile and target are modeled according to the Fermi-gas ansatz. The nucleons are represented by Gaussian shaped density distributions. The nucleon-nucleon interaction, implemented in the model, is based on a non-relativistic density-dependent Skyrme-type equation of state with additional Yukawa- and Coulomb potentials. Momentum dependent potentials are not used. The UrQMD collision term contains 55 different baryon species (including nucleon, delta and hyperon resonances with masses up to 2.25 GeV/c^2) and 32 different meson species (including strange meson resonances), which are supplemented by their corresponding anti-particle and all isospin-projected states. The states can either be produced in string decays, s-channel collisions or resonance decays. For excitations with higher masses than 2 GeV/c^2 a string picture is used. Full baryon/antibaryon symmetry is included. Elementary cross sections are fitted to available proton-proton or pion-proton data. Isospin symmetry is used when possible in order to reduce the number of individual cross sections which have to be parameterized or tabulated.

The UrQMD model was widely used in the current work for comparison with the NA49 data and predictions for the future CBM experiment.

Chapter 2

The NA49 experiment

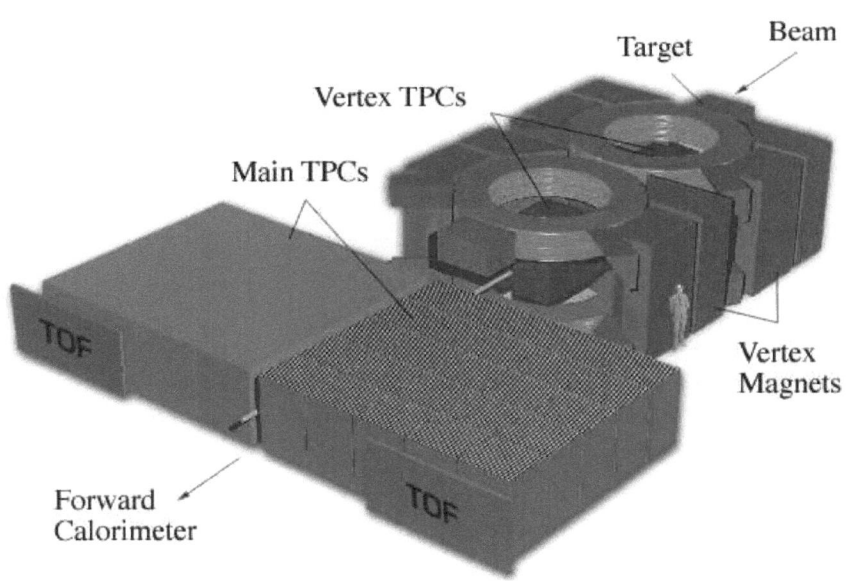

Figure 2.1: The NA49 detector.

The NA49 detector is a wide acceptance spectrometer for the study of hadron production in p + p, p + A, and A + A collisions at the CERN SPS [38]. The main components are 4 large-volume TPCs for tracking and particle identification via dE/dx.

TOF scintillator arrays complement particle identification. Calorimeters for transverse energy determination and triggering, a detector for centrality selection in p + A collisions, and beam definition detectors complete the set-up.

In the next section only those detectors are shortly described which are essential for the data analysis.

2.1 The magnets

Two super-conducting dipole magnets with a maximum combined bending power of 9 Tm at currents of 5000 A are centered on the beam line. Each has a width of 5700 mm and a length of 3600 mm. Their centers are positioned at about 2000 and 5800 mm from the target. The magnet yokes are configured in such a way that there is a maximum opening in the (horizontal) bending plane at the downstream end. A free gap of 1000 mm between the upper and lower coils leaves room for large- volume (TPC) tracking detectors. The coils have a central bore of 2000 mm in diameter and no pole tips are present. This causes field inhomogeneities with the minor components reaching 60% of the vertical component at the extremities of the active TPC volumes. The standard current settings for data taking corresponds to full field, nominally 1.5 T, in the first and reduced field, 1.1 T, in the second magnet.

2.2 The TPC tracking system

Time Projection Chamber (TPC) is a 3-D imaging chamber. Providing large volume and small amount of material, it is typically slow, with read-out time in the order of 50 μs. It was first proposed in 1976 and is used in many experiments.

2.2.1 Principles of operation

A TPC contains gas (for example Ar + 10-20 % CH_4), an electric field in the order of few 100 V/cm, a magnetic field, which should be as large as possible to measure momentum and to limit electron diffusion, and pad read-out to detect projected tracks [39].

The operation principle of a TPC with a charged track crossing the active volume is shown in figure 2.2. When passing through the detector gas a particle will produce primary ionization along its track. The y-coordinate is determined by measuring the drift time from the ionization event to the pad plane at the top. This is done using the

2.2. THE TPC TRACKING SYSTEM

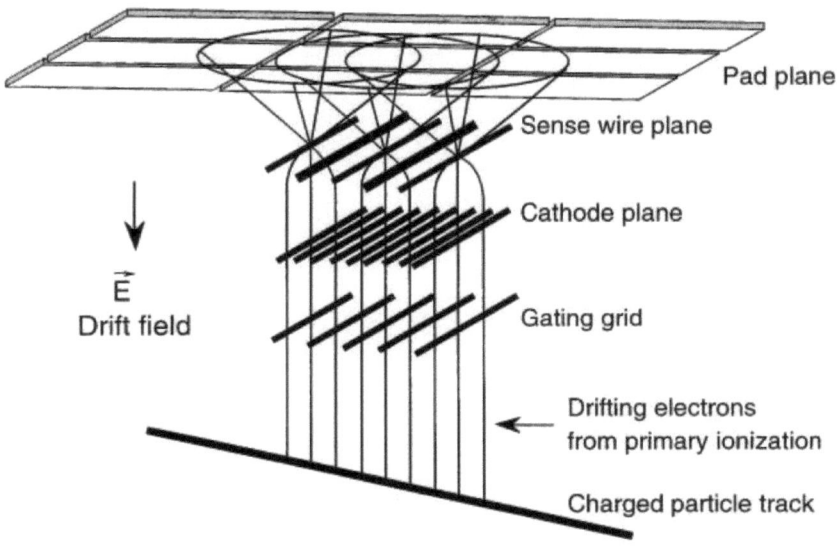

Figure 2.2: The scetch of the Time Projection Chamber (TPC) detector.

usual technique of a drift chamber. Pad read-out provides the measurement of x- and z-coordinates, thus track points are reconstructed in 3-D.

The TPC has just gas in the active volume, thus the material budget is small. The maximum drift lenght is 2 m which leads to long read-out time and thus slow data acquisition speed. Design requirements are typically high gas purity, uniform electric field and strong and uniform magnetic field. Particle identification is provided by the measurement of specific energy loss (dE/dx) which depends on the particle velocity and magnetic field. Using a TPC, the mass of the particle can be identified by measuring simultaniously its momentum and dE/dx.

High granularity of the pad read-out provides the possibility to operate the TPC in high track densities, which is relevant for heavy ion collisions at the CERN SPS energies.

2.2.2 Readout chambers

The NA49 TPC system deploys 62 readout proportional chambers each with a 72 × 72 cm² detector surface. The chambers are constructed using the classic design developed

	VTPC-1	VTPC-2	MTPC
Width	2 m	2 m	3.9 m
Length	2.5 m	2.5 m	3.9 m
Height	0.98 m	0.98 m	1.8 m
Gas	90% Ne, 10% CO_2	90% Ne, 10% CO_2	90% Ar, 5% CO_2, 5% CH_4
Sectors	6	6	25
Pad rows per sector	24	24	18
Pads per row	192	192	192/128
Pads	27648	27648	63360
Pad length	16/28 mm	28 mm	40 mm
Pad width	3.5 mm	3.5 mm	3.6/5.5 mm
Angle to beam	12 - 55	3 - 20	0/15

Table 2.1: Sizes and characteristics of the NA49 TPCs.

over many years of TPC technology. Seen from the drift space, a gating grid is followed by a cathode plane (Frisch grid) closing the proportional chamber volume made up by 20 μm sense wires interspaced with 125 μm field (zero potential) wires. Signals are picked up on the back cathode plane which is subdivided into pads.

The sizes, gas mixtures and characteristics of the NA49 TPCs are listed in table 2.1. There are two types of TPC detectors in NA49: two Vertex TPC's (VTPC1 and VTPC2) and two Main TPC' (left MTPC and right MTPC). Vertex TPC are smaller and are placed in the dipole magnetic field for measurement of particles' momentum.

2.3 Veto calorimeter

The centrality of Pb + Pb collisions is derived from the energy of the projectile spectators which is measured by Veto Calorimeter (VCAL). VCAL is located 26 m behind the target and consists of an electromagnetic section of a lead/scintillator sandwich of 16 radiation lengths which is followed by a hadron section of an iron/scintillator sandwich of 7.5 interaction lengths. Each section is divided into four segments. The light produced by the scintillators of each segment is transported by light guides located at the left and the right side of the calorimeter to a photo-multiplier. Due to this design the response of the

2.3. VETO CALORIMETER

calorimeter is strongly dependent on the position the particle traverse the scintillators. The light created in the center of the calorimeter has to go a longer way through the scintillator material and is weakened due to absorption or misses the photo-multipliers. This influences the resolution of the calorimeter.

The spectrum of total energy of the projectile spectators measured with VCAL of NA49 in Pb + Pb collisions at 158A GeV is shown in figure 2.3.

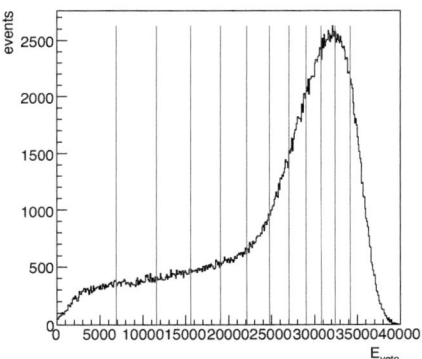

Figure 2.3: Distribution of the veto calorimeter energy for minimum bias Pb + Pb collisions at 158A GeV. Vertical lines show 5% centrality bins from 5 to 55%. The remaining 40% of events were triggered out and not recorded.

2.3.1 Centrality determination

The centrality of a reaction is the ratio of the cross-section σ_{trig} of the triggered reactions to the total inelastic cross-section σ_{inel}:

$$C = \frac{\sigma_{trig}}{\sigma_{inel}} \qquad (2.1)$$

The cross-section σ_{trig} can be estimated from the trigger rate of the experiment and the configuration of the target:

$$\sigma_{trig} = R_{trig} \cdot \left(\frac{M}{\rho \cdot d \cdot N_A}\right) \qquad (2.2)$$

where M is the molar mass of the target atoms, d is the thickness of the target and ρ is the target density. N_A is the Avogadro constant. One of the possibilities to determine the trigger rate is following formula:

$$R_{trig} = \left[\left(\frac{N_{trig}}{B_{gated}} \right)_{target\ in} - \left(\frac{N_{trig}}{B_{gated}} \right)_{target\ out} \right] \qquad (2.3)$$

where N_{trig} is the number of triggered events and B_{gated} is the number of beam particles which passed through the target within the same time interval, during which data acquisition system was idle.

Chapter 3

Data analysis

In this chapter the analysis of event-by-event particle ratio fluctuations in NA49 data will be explained. A list of the analysed data sets is shown in table 3.1. Here we assume that well measured momenta are availible.

Tag	Beam momentum, AGeV	Trigger	N_{ev}
20G+-20GeV-central-03A	20	central	360k
30G+-30GeV-central-02J	30	central	440k
1/4std+-40GeV-central-00W	40	central	420k
1/2std+-80GeV-central-01E	80	central	305k
std+-160GeV-central-00B	158	central	400k
std+-160GeV-central-256TB-01I	158	central	3M
std+-160GeV-minbias-low-int-01J	158	minbias	340k

Table 3.1: List of the analysed NA49 data sets.

The study of event-by-event fluctuations of particle yield ratios requires particle identification on an event-by-event basis. The NA49 experiment, with its particle identification via the dE/dx measurement, works in the relativistic rise region of the Bethe Bloch parametrisation and due to limited dE/dx resolution track-by-track identification is not possible. On the other hand, a standard fit of a single event dE/dx distribution will fail due to the low number of tracks per event (on average 60 tracks per event for $20A$ GeV beam energy). An alternative statistical method was proposed for this purpose [32, 42]. It is based on the maximum likelihood method and obtains particle yields per event using information of the dE/dx distribution of the whole data sample. This means, that before this statistical event-by-event fit, an inclusive fit has to be performed to obtain the parameters of the dE/dx distribution (for description of the parameters see section 3.1). These parameters are fixed later on and only relative yields are varied. Particle ratios in an event, which are needed for calculation of the fluctuations, are then calculated from the relative yields per event.

3.1 Inclusive fit

Particle identification by dE/dx measurement is based on the fact, that the specific energy loss of a particle depends on its velocity and is different for different particle species with the same momentum. The dependence of dE/dx on momentum is described by the Bethe-Bloch formula, and for NA49 by its parameteres (see figure 3.1). A simultaneous measurement of dE/dx and momentum provides particle identification [40]. In the current analysis, the dE/dx measured by the MTPCs of NA49 was used because it has higher quality and is more stable with respect to the dE/dx resolution, compared to measurements provided by the VTPCs. MTPC tracks also have more measurements along their trajectories and thus better dE/dx resolution (see section 2.2).

The inclusive dE/dx spectrum was divided into 2 bins in charge (+ and -), 20 bins in total momentum, 10 bins in transverse momentum and 8 bins in azimuthal angle. Table 3.2 shows the ranges and the bin sizes used in the current analysis.

Variable	Range	N_{bins}	Bin size
p_{tot}	1-40 and 2-120 GeV/c	20	logarithmic
p_t	0-2 GeV/c	10	0.2 GeV/c
ϕ	0-2π	8	0.25·π

Table 3.2: Binning of the phase space used in this analysis. Due to overlap of dE/dx distributions around momentum of 3 GeV/c first three bins in total momentum were skipped.

Figure 3.1 shows the transverse momentum and azimuthal angle integrated dE/dx distribution versus total momentum. Solid lines show the NA49 Bethe-Bloch parametrisation. The following momentum ranges were chosen: 1 - 40 GeV/c for beam energies 20A, 30A and 40A GeV and 2 - 120 GeV/c for 80A and 158A GeV. These momentum ranges are labeled as standard. Due to the overlap of the contributions around the total momentum of 3 GeV/c, particle species can be separated only above this momentum

3.1. INCLUSIVE FIT

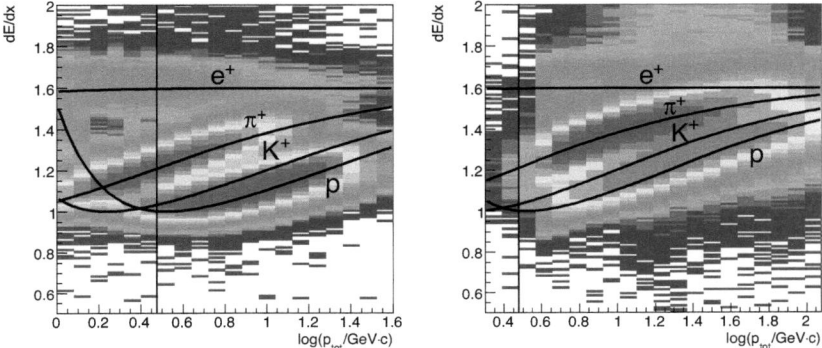

Figure 3.1: dE/dx as measured by the main-TPC as a function of total momentum for positive particles from central Pb + Pb collisions at $20A$ GeV (left picture) and at $158A$ GeV (right picture) beam energies. Solid lines are the NA49 Bethe-Bloch parametrisations. Vertical lines show momentum of 3 GeV/c.

value. Logarithmic binning for the total momentum axis and linear binning for the transverse momentum and the azimuthal angle were used.

The dE/dx distribution for tracks from the event ansamble was approximated by single particle distributions. Minimization of χ^2 was used as a fitting method. Each phase space bin was fitted separately. In order to exclude low statistics bins, 3000 entries were required in each distribution.

We have considered 4 particle species: electrons (positrons), pions, kaons and (anti-)protons. The distribution for each particle was considered to be Gaussian. All Gaussians are characterized by their position (relative to the pion position), their amplitude and a position dependent width. So there are 9 free parameters in the fit. Summarizing this, the dE/dx fitting function e.g. for particles with positive charge is given by:

$$f(d) = A_{e^+} exp\bigg[-0.5\Big(\frac{d - y_{e^+}(\frac{dE}{dx})_{\pi^+}}{q_{e^+}}\Big)^2\bigg]$$
$$+ A_{\pi^+} exp\bigg[-0.5\Big(\frac{d - (\frac{dE}{dx})_{\pi^+}}{q_{\pi^+}}\Big)^2\bigg]$$
$$+ A_{K^+} exp\bigg[-0.5\Big(\frac{d - y_{K^+}(\frac{dE}{dx})_{\pi^+}}{q_{K^+}}\Big)^2\bigg]$$

$$+ A_p exp\left[-0.5\left(\frac{d - y_p(\frac{dE}{dx})_{\pi^+}}{q_p}\right)^2\right], \tag{3.1}$$

where A_{e^+}, A_{π^+}, A_{K^+}, A_p - are the peak amplitudes; $y_\alpha = (\frac{dE}{dx})_\alpha/(\frac{dE}{dx})_{\pi^+}$ - is the relative position of particle type α and $q_{alpha} = \sigma y_\alpha (\frac{dE}{dx})_{\pi^+}$ is the width. The same fitting function was used for all phase space bins. The peak positions were initialized from the Bethe-Bloch parametrisation of NA49 and for the first iteration were fixed to this initial value, to obtain an estimate of the amplitudes and the width. In the second iteration all 9 parameters were free.

Dedicated tools for visualization of the fitting procedure were developed, which allow to control it on-line bin by bin. Several examples are shown for central Pb + Pb collisions at 20A GeV beam energy for positive particles with p$_{tot}$=3.3 GeV/c (figure 3.2), p$_{tot}$=14.6 GeV/c (figure 3.4), p$_{tot}$=8.4 GeV/c (figure 3.3) and at 158A GeV beam energy for positive particles with p$_{tot}$=4.1 GeV/c (figure 3.5), p$_{tot}$=7.6 GeV/c (figure 3.6) and p$_{tot}$=88.7 GeV/c (figure 3.7). Loose set of track cuts was used (see section 3.4).

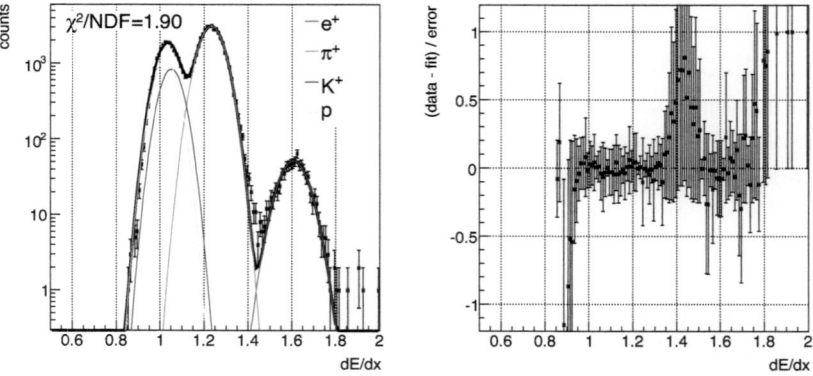

Figure 3.2: Visualization of the inclusive dE/dx fit (left picture) for central Pb + Pb collisions at 20A GeV beam energy for positive particles with p$_{tot}$=3.3 GeV/c, p$_t$=0.3 GeV/c and ϕ=0.393 rad. The right picture shows the deviation of the fit function from the data distribution relative to the data error.

The performance of the fit expressed in terms of χ^2 distribution as function of p$_t$ and log(p$_{tot}$) is shown in figure 3.8 for central Pb + Pb collisions at 20A and 158A GeV beam energies, positive tracks and the first bin of the azimuthal angle. It can be seen that the

3.1. INCLUSIVE FIT

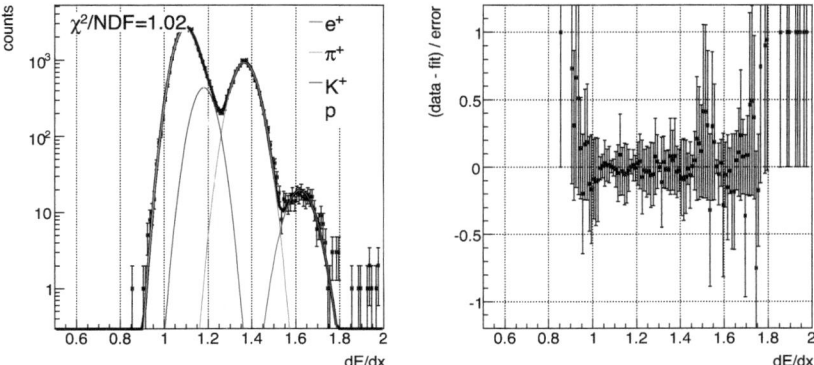

Figure 3.3: Visualization of the inclusive dE/dx fit (left picture) for central Pb + Pb collisions at $20A$ GeV beam energy for positive particles with p_{tot}=8.4 GeV/c, p_t=0.3 GeV/c and ϕ=0.393 rad. The right picture shows the deviation of the fit function from the data distribution relative to the data error.

overall performance of the inclusive fit is satisfactory but it is worse for the higher beam energy due to the fact that the systematic errors become larger than the statistical ones.

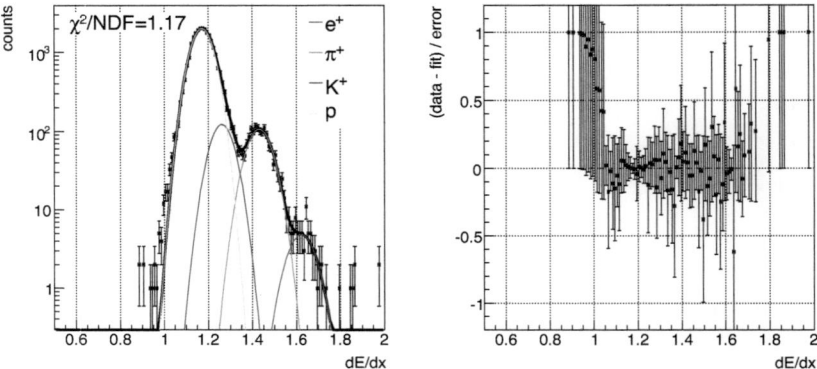

Figure 3.4: Visualization of the inclusive dE/dx fit (left picture) for central Pb + Pb collisions at $20A$ GeV beam energy for positive particles with p_{tot}=14.6 GeV/c, p_t=0.3 GeV/c and ϕ=0.393 rad. The right picture shows the deviation of the fit function from the data distribution relative to the data error.

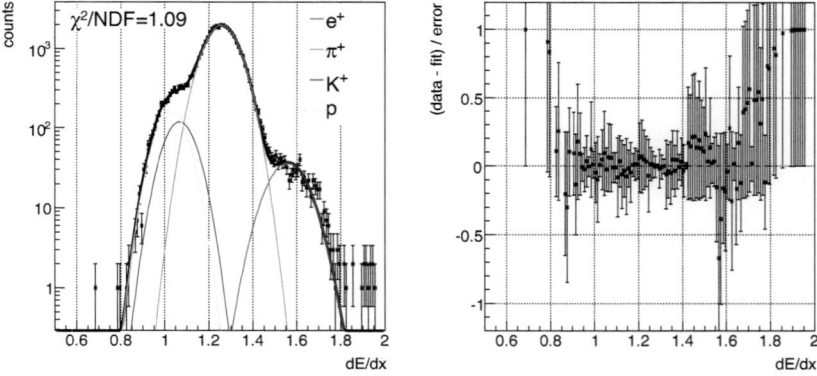

Figure 3.5: Visualization of the inclusive dE/dx fit (left picture) for central Pb + Pb collisions at $158A$ GeV beam energy for positive particles with p_{tot}=4.1 GeV/c, p_t=0.3 GeV/c and ϕ=0.393 rad. The right picture shows the deviation of the fit function from the data distribution relative to the data error.

3.1. INCLUSIVE FIT

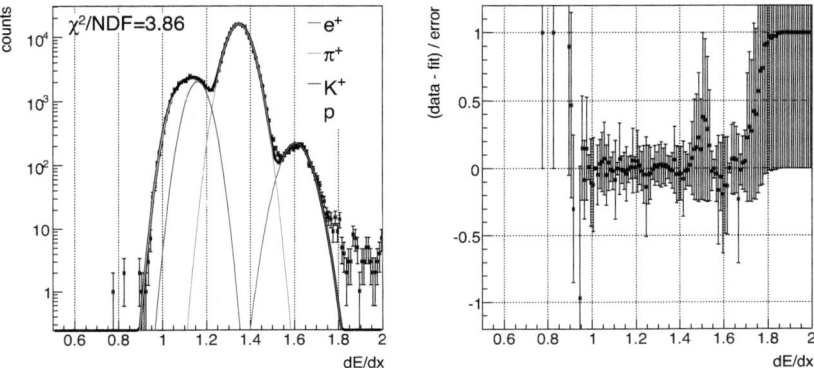

Figure 3.6: Visualization of the inclusive dE/dx fit (left picture) for central Pb + Pb collisions at 158A GeV beam energy for positive particles with p_{tot}=7.6 GeV/c, p_t=0.3 GeV/c and ϕ=0.393 rad. The right picture shows the deviation of the fit function from the data distribution relative to the data error.

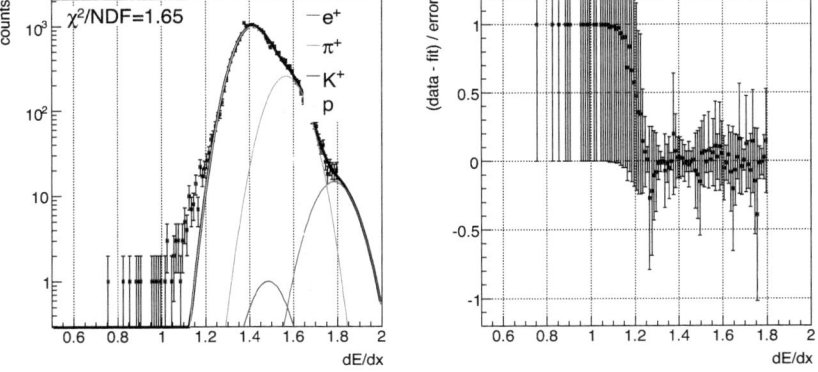

Figure 3.7: Visualization of the inclusive dE/dx fit (left picture) for central Pb + Pb collisions at 158A GeV beam energy for positive particles with p_{tot}=88.7 GeV/c, p_t=0.3 GeV/c and ϕ=0.393 rad. The right picture shows the deviation of the fit function from the data distribution relative to the data error.

28   CHAPTER 3. DATA ANALYSIS

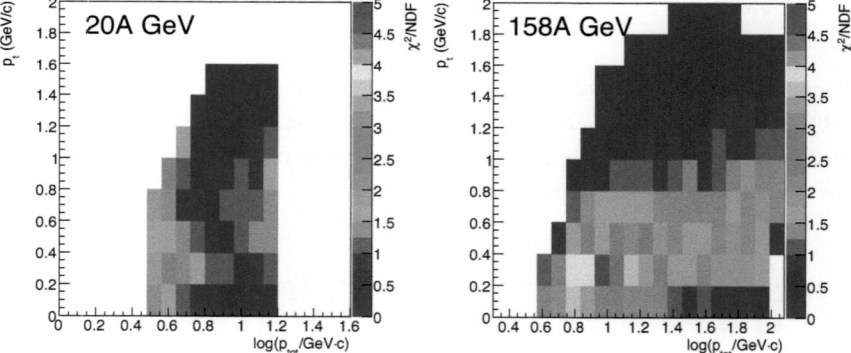

Figure 3.8: Performance of the inclusive fit for central Pb + Pb collisions at 20A GeV (left picture) and at 158A GeV (right picture) beam energies. The scales of the z-axes are the same on both plots.

3.1. INCLUSIVE FIT

The absolute positions of the Gaussians, fitted during the second iteration, are shown in figure 3.9. Each point represents one phase space bin for positive particles. The solid lines show the Bethe-Bloch parametrisation of NA49. One can observe a spread of points around the true value which indicates the variation of the peak positions from bin to bin. It is especially big for positrons (electrons), most probably, due to low statistics and lower quality (less length, odd angles, secondaries).

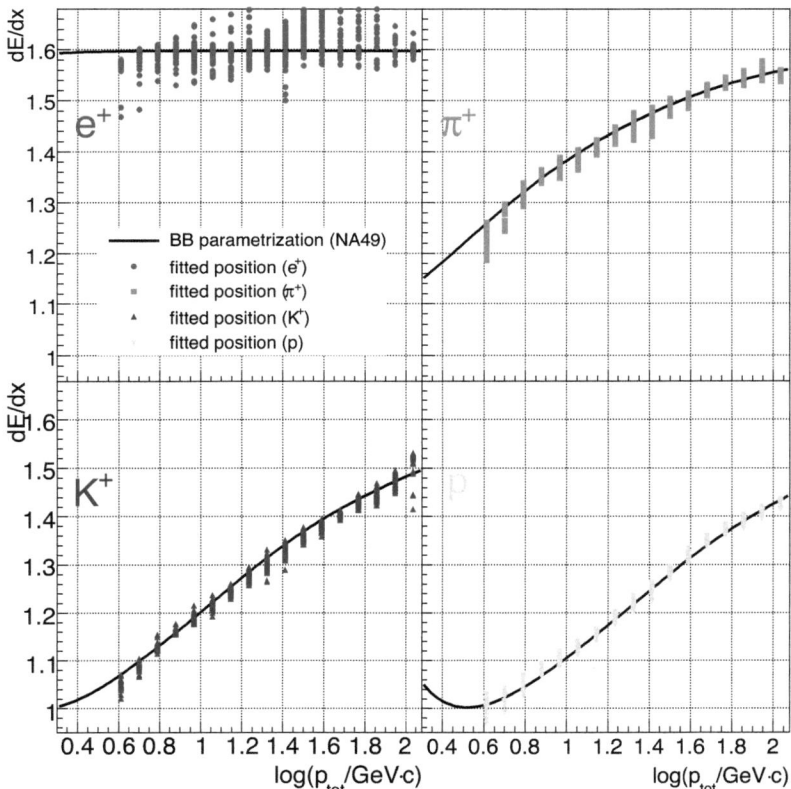

Figure 3.9: Positions of the Gaussians as a function of total momentum for positive particles from central Pb + Pb collisions at 158A GeV beam energy. Different p_t and ϕ bins are shown as dots. Solid lines show the Bethe-Bloch parametrisation of NA49.

The extracted amplitudes for positive particles from the first bin of the azimuthal angle as a function of total and transverse momentum are shown in figure 3.10.

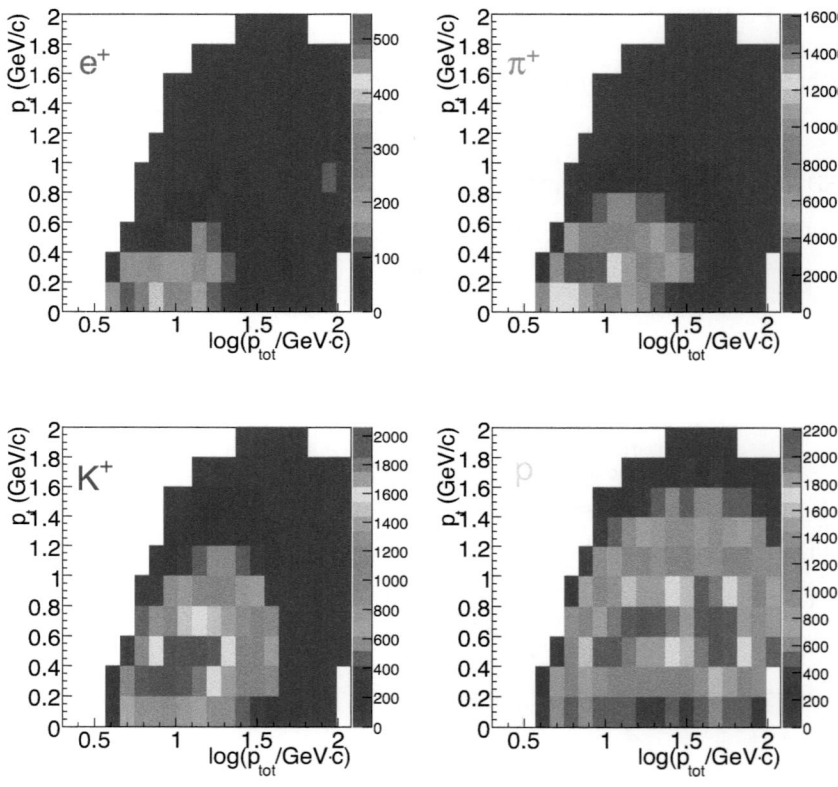

Figure 3.10: Amplitudes as a function of total and transverse momentum for positive particles from central Pb + Pb collisions at 158A GeV beam energy.

3.1. INCLUSIVE FIT

For a crosscheck of the extracted amplitudes they were transformed into raw yield distributions as function of transverse momentum and rapidity and compared to the fully corrected dn/dy spectrum of K^+ from central Pb + Pb collisions at 158A GeV, published by NA49 [41]. The azimuthal angle wedge of $\pm 45°$ was used, in which the acceptance is close to 100%. A simple factor of 4 (90° out of 360°), which corrects for the selected ϕ coverage, was introduced. Since no extrapolation in p_t was done in the current analysis, one expects agreement of both spectra (with 5% relative difference due to resonance feeddown) in the rapidity range from 0.5 to 1, while at higher rapidities they might disagree due to limited p_t coverage (see figure 3.12). From figure 3.11 one can conclude satisfactory agreement of ϕ-integrated dn/dy yields of K^+ at 158A GeV beam energy in the rapidity range from 0.5 - 1.

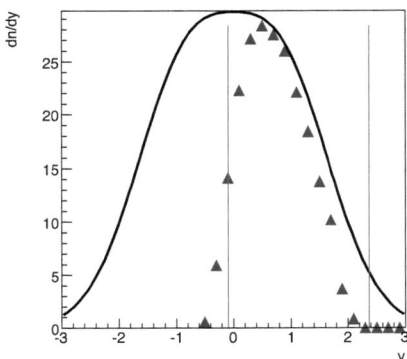

Figure 3.11: Comparison of raw K^+ dn/dy spectra obtained by the current analysis (points) and published by NA49 (curve) for central Pb + Pb collisions at 158A GeV beam energy. Two vertical lines indicate the kinematical range of the current analysis and show the rapidity values in the CM frame for p_{tot}=3 GeV/c, p_t=0 GeV/c (left line) and p_{tot}=120 GeV/c, p_t=1 GeV/c (right line).

The phase space coverage (p_t versus rapidity distributions) for positive particles from central Pb + Pb collisions at 158A GeV beam energy is shown in figure 3.12. Information on the acceptance of the NA49 detector for the complete CERN SPS energy range can be found in [32].

3.1. INCLUSIVE FIT

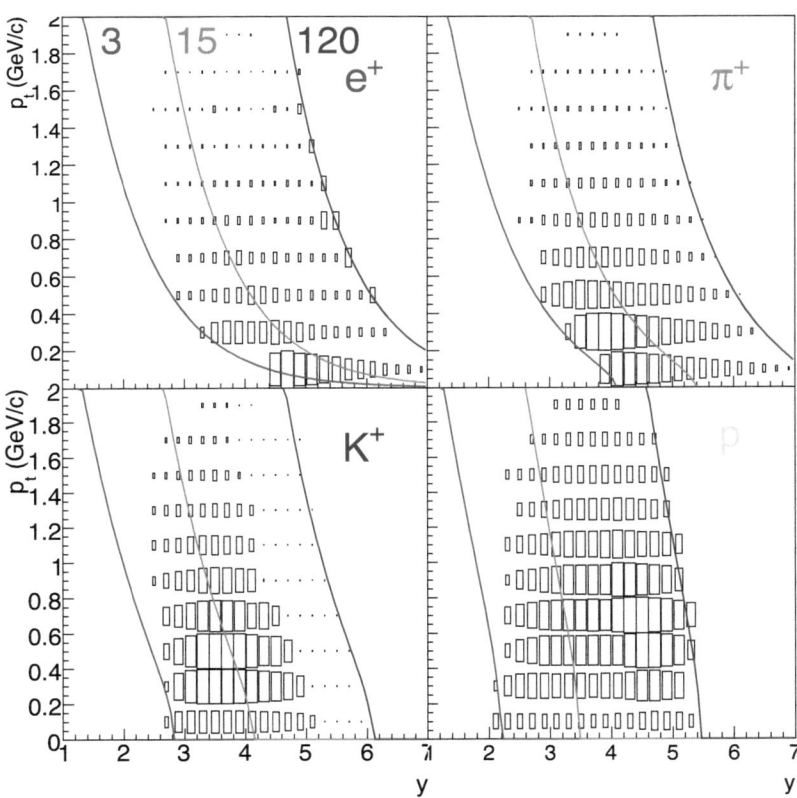

Figure 3.12: Phase space coverage for positive particles from central Pb + Pb collisions at 158A GeV beam energy. Solid lines show the momentum cuts used in the current analysis: lower momentum cut of 3 GeV/c (left line) and two different higher momentum cuts of 15 GeV/c (middle line) and 120 GeV/c (right line).

The width of the dE/dx distribution of each particle was parametrized as follows:

$$\sigma = a\left(\frac{dE}{dx}\right)^{\alpha}, \tag{3.2}$$

where the parameter α depends on the gas of the TPC. For the MTPCs of NA49 it is equal to 0.65 [32]; a is the relative width of the peaks and is assumed to be common for all particles. In fact, the parameter a depends on $1/\sqrt{n}$ with n being the number of dE/dx measurements per track, thus is larger for shorter tracks. This dependence was neglected in the current analysis, since close to all tracks accepted in the MTPCs are long (number of dE/dx measurements is close to 100). The fitted common relative width as a function of total and transverse momentum for positive particles from the first bin of azimuthal angle for central Pb + Pb collisions at 158A GeV beam energy is shown in figure 3.13. A dE/dx resolution on the order of 4% - 4.5% is obtained.

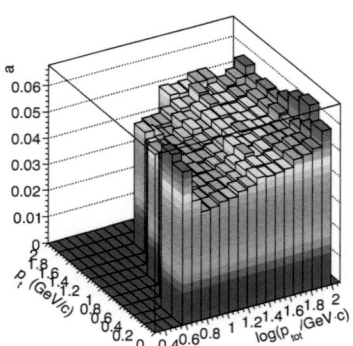

Figure 3.13: dE/dx resolution as a function of total and transverse momentum for central Pb + Pb collisions at 158A GeV beam energy.

The stability of the inclusive fit with respect to different centralities of Pb + Pb collisions at 158A GeV beam energy is shown in figure 3.14. The peak positions are very stable while there are small variations of the absolute width.

3.2. EVENT-BY-EVENT FIT

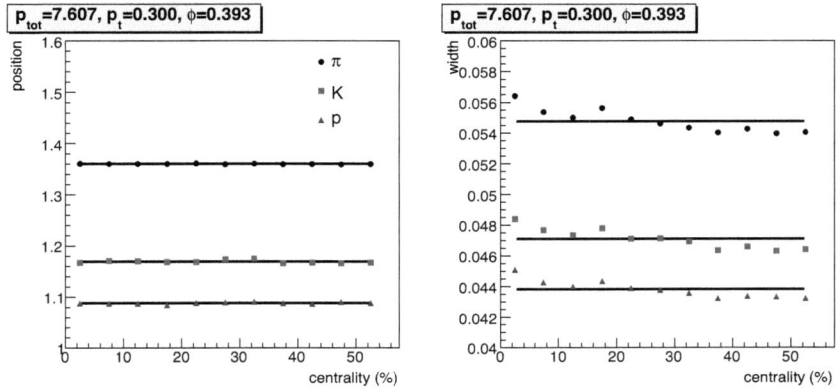

Figure 3.14: Stability of the inclusive fit with respect to different centralities of Pb + Pb collisions at 158A GeV beam energy for p$_{tot}$=7.6 GeV/c, p$_t$=0.3 GeV/c and ϕ= 0.393 rad.

3.2 Event-by-event fit

Results from the inclusive fit are filled into a look up table, which consists of 9 fit parameters for each phase space bin. These Probability Density Functions (PDFs) were used in the further analysis to obtain event-by-event particle yields and are shown in figure 3.15.

As already pointed out in section 3.1 there are too few entries in the dE/dx distribution per event if divided in bins of different momenta, the standard procedure with minimization of the χ^2 fails. Thus the Maximum Likelihood Method (MLM) was used to extract event-wise yields [42]. Using the PDFs one can define the probability for each track to be one of the particle types. The probabilities for each track are weighted by the relative yield of each particle and summed up. Multiplying this variable for all tracks per event, a likelihood value for a certain yield distribution is obtained. This likelihood is maximized eventwise in the fit. So, the likelihood value is given by:

$$L(\{\theta_\alpha\}) = \prod_{i=1}^{N} \sum_\alpha \theta_\alpha f_\alpha(q^i, p_{tot}^i, p_t^i, \phi^i, (dE/dx)^i), \quad (3.3)$$

where $\alpha = \{e, \pi, K, p\}$, θ_α is the relative yield of the particle α per event, N is number of tracks in an event, and $f_\alpha(...)$ is the PDF for particle α. The function is constructed such that:

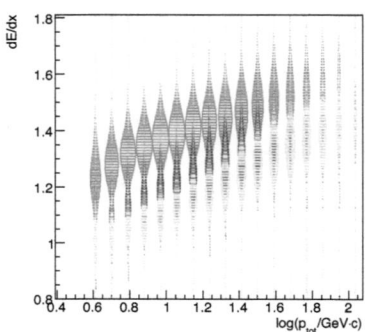

Figure 3.15: Probability Density Functions for positively charged particles from central Pb + Pb collisions at 158A GeV beam energy.

$$\sum_\alpha \theta_\alpha = 1 \qquad (3.4)$$

The sum of all relative yields per event is unity.

The set of relative yields $\{\theta_\alpha\}^*$ for which $L(\{\theta_\alpha\}^*)$ becomes maximum is the best estimate of true yields per event. The fit was done in two steps: first the positron (electron) and the kaon yields are fixed and the proton yield is varied. During the second step the kaon yield is varied and the rest of the yields are fixed. Further steps with adjusted input parameters were done as a systematic check and no significant variation of the results was observed. The pion yield is extracted from the constraint as given in equation 3.4. The particle yield ratio in an event is then given by the ratio of relative yields obtained for this event.

From the technical point of view it is more convinient to minimize the following function:

$$l = -ln(L) \qquad (3.5)$$

The Minuit package was used for the minimization procedure. The dependence of function l from equation 3.5 on the relative kaon and proton yields for a single event is shown in figure 3.16. As a crosscheck of the minimization with Minuit, a manual minimization was performed by a two-dimensional loop over the relative kaon and proton yields. The position of the manually obtained minimum was found to be in agreement

3.3. EXTRACTION OF DYNAMICAL FLUCTUATIONS

with the Minuit results (see figure 3.16).

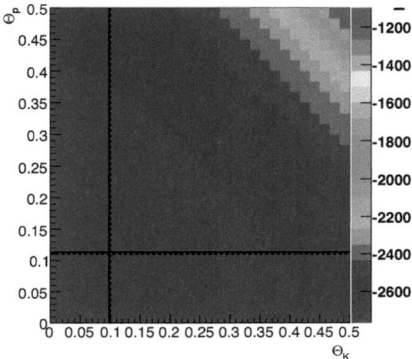

Figure 3.16: Dependence of the function l (equation 3.5) on the relative kaon and proton yields for central Pb + Pb collisions at 158A GeV beam energy. Solid lines show the position of the minimum found by Minuit, while the dashed lines indicate the manually found minimum.

In case of a low number of tracks per event problems are encountered when applying the maximum likelihood method. One of them is the presence of events with zero number of, for example, kaons. In this case the eventwise kaon to pion ratio distribution develops a spike at zero, which influences the calculation of the fluctuations. For more details on the subject of spike at zero see section 5.1.2. Systematic checks of the MLM were performed for integer and negative representation of the yields in the event-by-event fit. It turns out that the event-by-event fit with the integer particle yields per event and/or with negative yields is not possible with the Minuit. The fit with integers has to be done manually and does not result in difference of dynamical fluctuations, while the fit with negative yields is completely unstable.

3.3 Extraction of dynamical fluctuations

Dynamical particle ratio fluctuations are defined as follows. Let N_π and N_K denote the numbers of measured charged pions and kaons, respectively, in an event, then $\frac{N_K}{N_\pi}$ is the kaon to pion ratio for this event. By measuring a set of events, we can numerically calculate the mean and RMS for the distribution of this event-by-event particle ratio and define the corresponding event-by-event fluctuations as:

$$\sigma = \frac{RMS}{MEAN} \qquad (3.6)$$

The statistical error of this variable is, neglecting the error in the mean,

$$\delta(\sigma) = \frac{\sigma}{\sqrt{2N_{ev}}}, \qquad (3.7)$$

where N_{ev} is the number of analysed events.

Defined in this way, σ contains a non-dynamical contribution from finite-number statistics and detector effects (resolution, acceptance) and a "dynamical" contribution which may be connected with critical phenomena:

$$\sigma^2 = \sigma_{stat}^2 + \sigma_{dyn}^2, \qquad (3.8)$$

The non-dynamical contribution can experimentally be determined by a careful event mixing technique. The idea of event mixing is to destroy the correlations between particles which are present in the data events and in this way to estimate pure statistical fluctuations. During event mixing tracks are randomly drawn from the real events, each time from a different one. Thus the size of the event pool has to be larger than the maximum number of tracks per event. Only events of the same centrality class (see section 2.3.1) are used. An additional constraint is the total multiplicity of registered charged particles in the acceptance which is kept the same as in the real events. In this way mean values of the eventwise ratio distributions from data and mixed events are as close to each other as possible. One allows for a difference of mean values due to correlations between particles which are present in the data events and removed from mixed events by construction. After applying the described procedure one can determine the fluctuations of the particle yield ratio in the mixed events, the so called σ_{mix}. The dynamical fluctuations are then defined as

$$\begin{aligned} \sigma > \sigma_{mix} &: \quad \sigma_{dyn} = \sqrt{\sigma^2 - \sigma_{mix}^2}, \\ \sigma < \sigma_{mix} &: \quad \sigma_{dyn} = -\sqrt{\sigma_{mix}^2 - \sigma^2}, \end{aligned} \qquad (3.9)$$

where the first equation corresponds to an anti-correlation (broadening of the distribution compared to the background) and the second to a correlation (narrowing). The statistical error of this quantity is obtained by error propagation and equation 3.7 as:

$$\delta(\sigma_{dyn}) = \frac{1}{|\sigma_{dyn}|\sqrt{2N_{ev}}} \sqrt{\sigma^4 + \sigma_{mix}^4} \qquad (3.10)$$

3.4 Event and track cuts

To increase the quality of the data, one introduces certain event and track cuts. The following event cuts have been used in this work. Values of these cuts depend on the specific analysis task. They will be given in chapter 4. Centrality selection by E_{veto} binning; rejection of events with failed dE/dx evaluation ($\sum dE/dx = 0$); number of tracks used for the event-by-event fit is larger than N_{min}; multiplicity cut, for rejection of outlier events. Track cuts can be varied in order to either select only a small set of highest quality tracks or to allow for more tracks but with less strict requirements. Typically, a tight and loose set of cuts was used in order to evaluate systematic errors of the results (see section 3.5). The loose set of track cuts is described as follows: number of found points is larger than 30 (MTPCs); number of potential points is larger than 0 (in at least one VTPC and one MTPC); at least 50% of potential points found (in MTPC's); dE/dx value is smaller than 1.8; track is in one of the p_{tot}, p_t, ϕ bins, which was successfully fitted; beam rapidity cut: $(y_{proton} < (y_{beam} - 1))$ or $((y_{proton} > (y_{beam} - 1))$ and $(p_t > 0.2))$.

3.5 Systematic errors due to track selection

The tight set of track cuts, used for evaluation of systematic errors, consists of following track cuts: number of found points in VTPC1 is larger than 10; number of found points in VTPC2 is larger than 10; number of found points in MTPCs is larger than 30; at least 50% of potential points found; track is fitted to the primary vertex (iflag equal 0); x component of the impact parameter is smaller than 4 cm; y component of the impact parameter is smaller than 0.5 cm; dE/dx value is smaller than 1.8; track is in the fitted bin of the dE/dx container; beam rapidity cut: $(y_{proton} < (y_{beam} - 1))$ or $((y_{proton} > (y_{beam} - 1))$ and $(p_t > 0.2))$.

Chapter 4

Results of the data analysis

4.1 Energy dependence

Before results of the analysis of the centrality dependence of dynamical particle ratio fluctuations will be discussed, one would first like to validate and verify the software routines, which were re-implemented, comparing to the published NA49 results on the energy dependence of the fluctuation signal [32]. In this section the energy dependence of dynamical fluctuations of the K/π, p/π and K/p ratios and their comparison to previously obtained results from an independent analysis will be presented. The description of the background estimation (statistical fluctuations) and the extraction of the dynamical fluctuations is described in section 3.3. Values of the E_{veto} and multiplicity cuts which have been used for centrality selection and outlier event removal are listed in table 4.1.

Beam energy, AGeV	E_{veto} max	N_{acc} min	N_{acc} max
20	800	45	85
30	1350	80	150
40	1230	120	200
80	9800	260	380
158	8370	460	660

Table 4.1: Values of the E_{veto} and multiplicity cuts on accepted particles N_{acc} used in the analysis of the energy dependence.

As example for the centrality selection used in this analysis the distribution of the energy deposited in the Zero Degree Calorimeter of NA49 for 7% most central Pb + Pb collisions at $20A$ GeV beam energy is shown in figure 4.1.

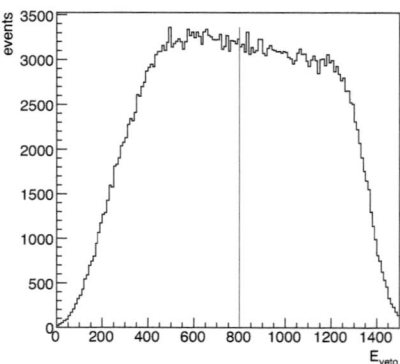

Figure 4.1: Distribution of the energy measured by the Zero Degree Calorimeter of NA49 for central Pb + Pb collisions at $20A$ GeV beam energy. The vertical line shows the cut value (3.5% most central collisions) used for this data set.

Figure 4.2 shows the distributions of the event-by-event kaon to pion yield ratio for data and mixed events in central Pb + Pb collisions at all beam energies available for NA49. The ratio distributions become broader with lower beam energy due to finite number statistics and finally run into zero. A spike at zero for lowest SPS energies appears. Its effect on the dynamical fluctuations will be discussed later. The dynamical fluctuations are positive due to two reasons as will be discussed in chapter 5: anti-correlation between K and π and K^+-K^- correlations due to ϕ decay.

The relative widths of data and mixed events distributions together with the dynamical fluctuations are given in the legends. As expected statistical fluctuations decrease with increasing beam energy due to higher particles multiplicities.

4.1. ENERGY DEPENDENCE

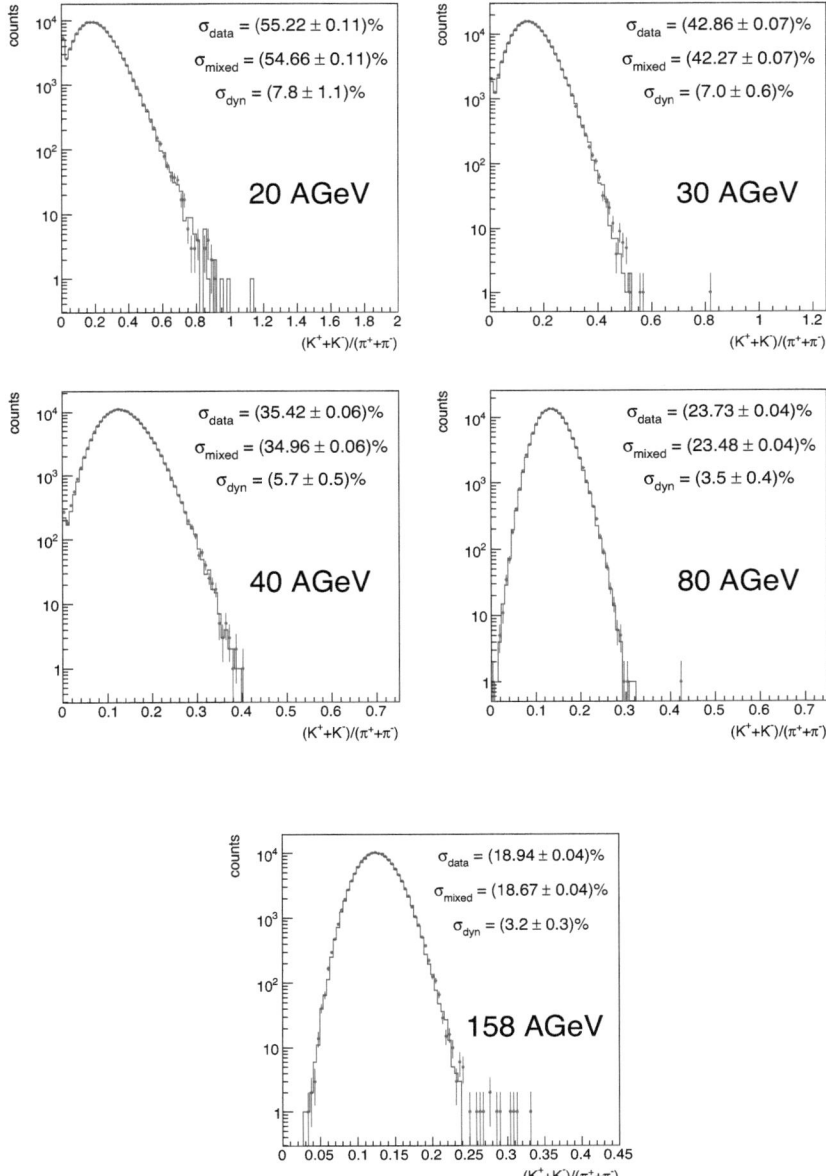

Figure 4.2: Distributions of the event-by-event K/π ratio for data (points) and mixed (histogram) events for 3.5% most central Pb + Pb collisions at different beam energies.

The center of mass energy dependence of the dynamical fluctuations of the kaon to pion yield ratio is shown in figure 4.3. Systematic errors were estimated using the variation of track cuts (see section 3.4) and are shown as the gray band. An increase of the fluctuations towards lower SPS energies is observed. The results are shown for the standard and limited total momentum ranges. The values of the dynamical fluctuations are in a good agreement with the published NA49 results, for which the momentum range from 3 - 40 (120) GeV/c was used [32].

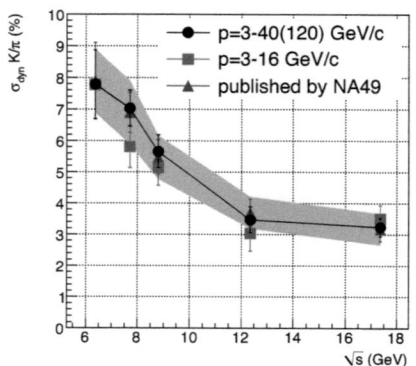

Figure 4.3: Energy dependence of the dynamical fluctuations of the kaon to pion ratio for the 3.5% most central Pb + Pb collisions. Gray band shows the estimated systematic error. The following momentum ranges were considered: 3 - 40 GeV/c for 20A, 30A, 40A GeV and 3 - 120 GeV/c for 80A and 158A GeV beam energies.

The distributions of the event-by-event proton to pion ratio for data and mixed events for all measured NA49 energies is shown in figure 4.4. The relative widths of data and mixed events distributions together with the dynamical fluctuations are given in the legends. As expected statistical fluctuations decrease with increasing beam energy due to higher particles multiplicities. The extracted dynamical fluctuations are negative, which indicates a strong correlation between protons and pions present in the data. This is most probably due to delta resonance decay. More details towards this hypothesis will be discussed in chapter 5.

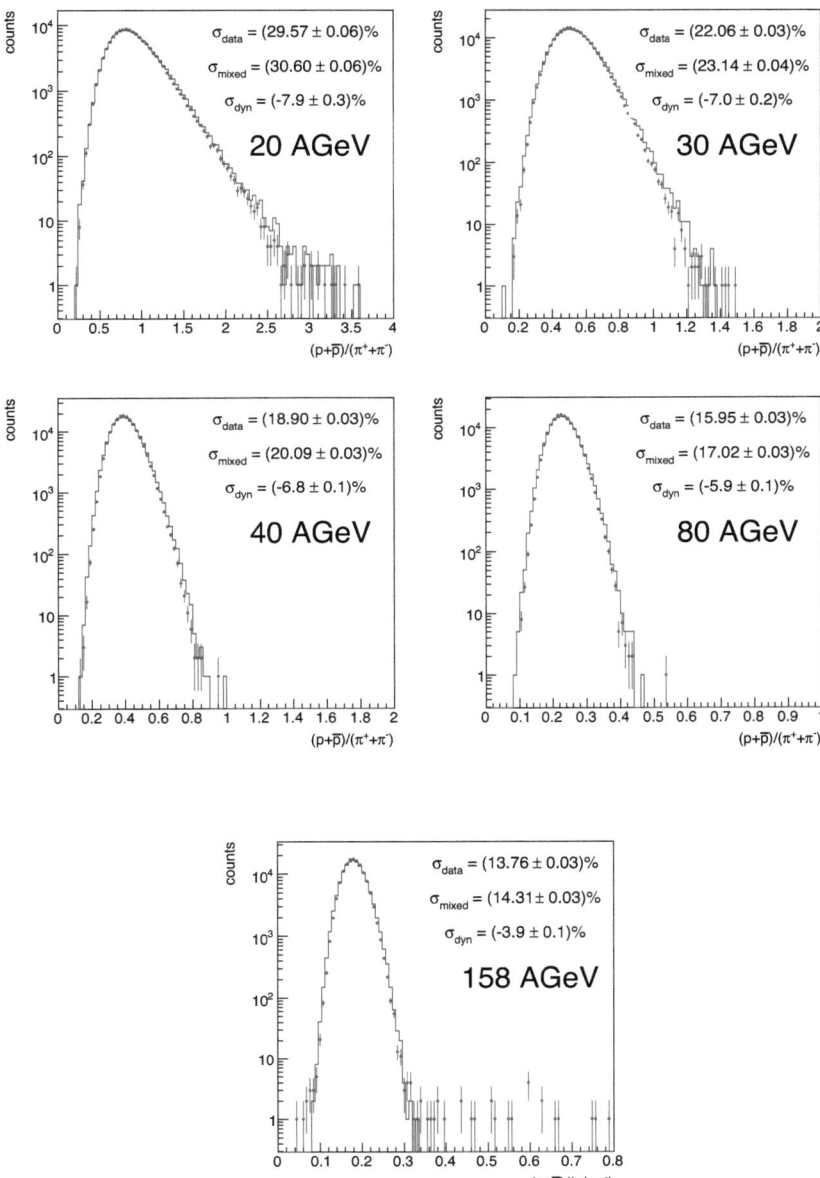

Figure 4.4: Distributions of the event-by-event proton to pion ratio for data (points) and mixed (histogram) events for the 3.5% most central Pb + Pb collisions at different beam energies.

The dynamical fluctuations of the proton to pion yield ratio as a function of center of mass energy are shown in figure 4.5. We observe a 1% difference of the results for the standard momentum range with that one published by NA49 for the top SPS energy [32]. Note, that published values are the average between the results for tight and loose track cuts, while our results are for the loose set of track cuts. The same data set was analysed using the same cut on centrality. A different analysis [43] also results in 1% difference as compared to the published value.

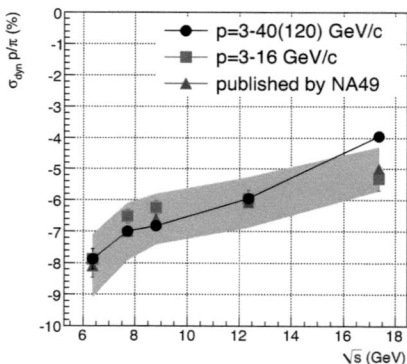

Figure 4.5: Energy dependence of the dynamical fluctuations of the proton to pion ratio for the 3.5% most central Pb + Pb collisions. The estimated systematic error is shown by the gray band.

Figure 4.6 shows the distributions of the event-by-event kaon to proton yield ratios for data and mixed events for the beam energies $20A$ - $158A$ GeV. The fluctuations change their sign from positive (for the lowest SPS energy) to negative (for the $30A$, $40A$, $80A$ and $158A$ GeV).

4.1. ENERGY DEPENDENCE

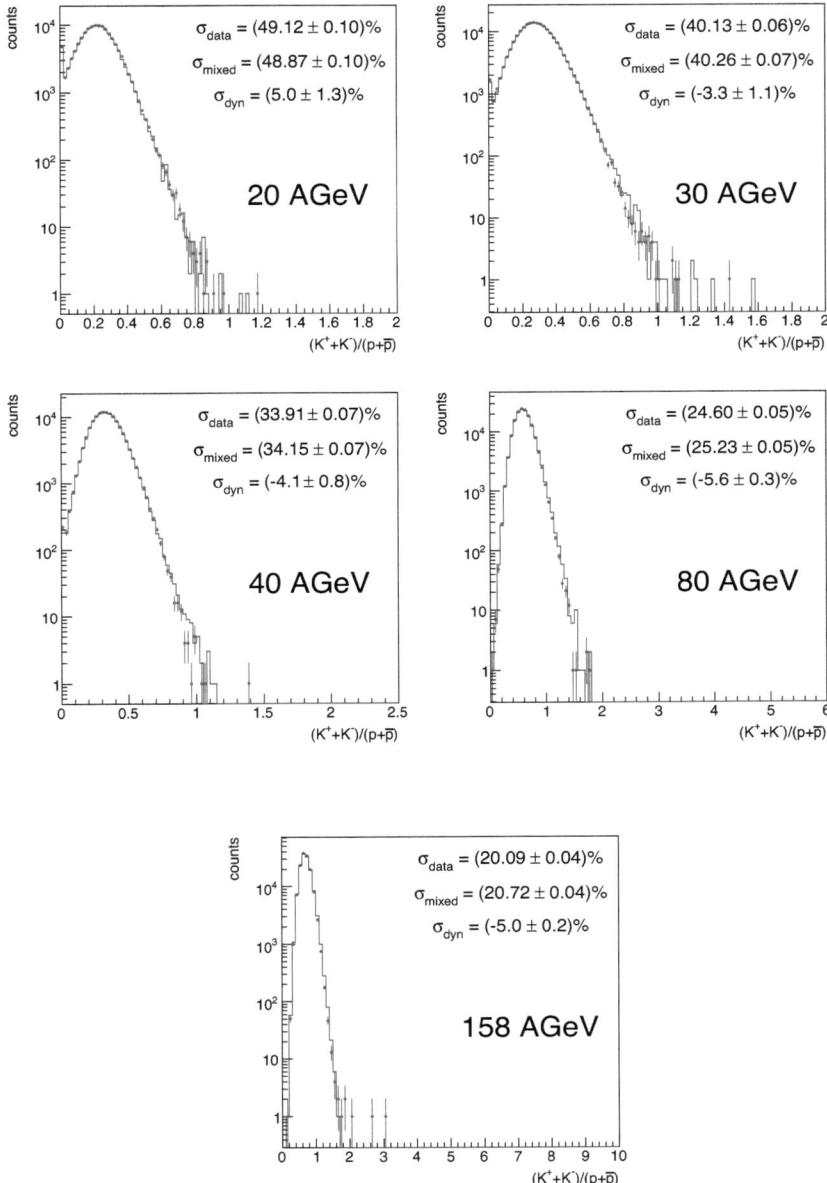

Figure 4.6: Distributions of the event-by-event kaon to proton ratio for data (points) and mixed (histogram) events for the 3.5% most central Pb + Pb collisions at different beam energies.

The energy dependence of the dynamical fluctuations of the kaon to proton yield ratio is shown in figure 4.7. The results on the K/p fluctuations are in agreement with a recent analysis presented in [43].

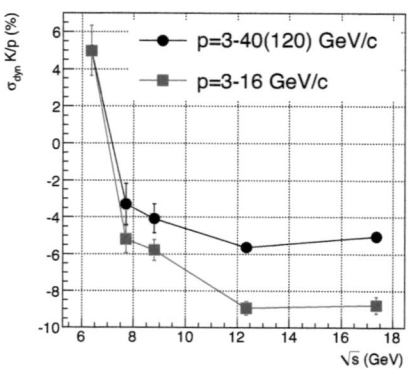

Figure 4.7: Energy dependence of the dynamical fluctuations of the kaon to proton ratio for the 3.5% most central Pb + Pb collisions.

In contrast to the kaon to pion and proton to pion ratios, the dynamical fluctuations of the kaon to proton yield ratio is more sensitive to the selected momentum range at higher beam energies. More studies on this subject will be presented in [43].

4.2 Dependence on centrality bin size

So far the paradigm was to use as small a centrality bin size as possible to reduce fluctuations from centrality. However, the particle yield ratios should be volume independent to first order. Before investigating the centrality dependence of particle ratio fluctuations the effect of the centrality bin size on particle ratio fluctuations has therefore to be studied for 158A GeV beam energy (alternative would be to use minimum bias data set at 40A GeV, but most probably particles multiplicities are too small to analyze particle ratio fluctuations). This investigation should help to select a proper centrality bin size, which does not introduce a bias to the results and allows to use a maximum on statistics. Thus we consider the study of the influence of the centrality bin size as the first step towards the centrality dependence of particle yield ratio fluctuations from NA49 data. Both central and semi-peripheral (starting from centrality of 30%) events were considered. The

4.2. DEPENDENCE ON CENTRALITY BIN SIZE

following bin sizes were studied: (0-3)%, (0-3.5)%, (0-5)%, (0-10)%, (0-15)%, (0-17.5)% and (0-20)% as well as (30-33)%, (30-33.5)%, (30-35)%, (30-40)%, (30-45)%, (30-47.5)% and (30-50)%.

4.2.1 Central events

The measured mean yields per event in the acceptance used for event-by-event fits for pions, kaons and protons as function of centrality bin size are shown in figure 4.8. An almost linear decrease of the yields (except first several points) with increasing bin size is observed.

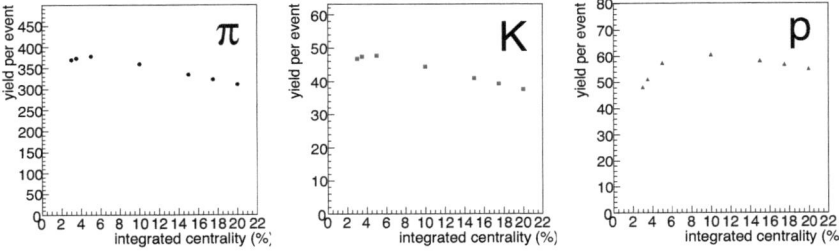

Figure 4.8: Mean particle multiplicities as extracted from the event-by-event fit as a function of centrality bin width for central Pb + Pb collisions at 158A GeV (data set 01J, see table 3.1). From left to right: pions, kaons, protons.

The dependence of the mean K/π and p/π ratios on the centrality bin size is shown in figure 4.9.

50   CHAPTER 4. RESULTS OF THE DATA ANALYSIS

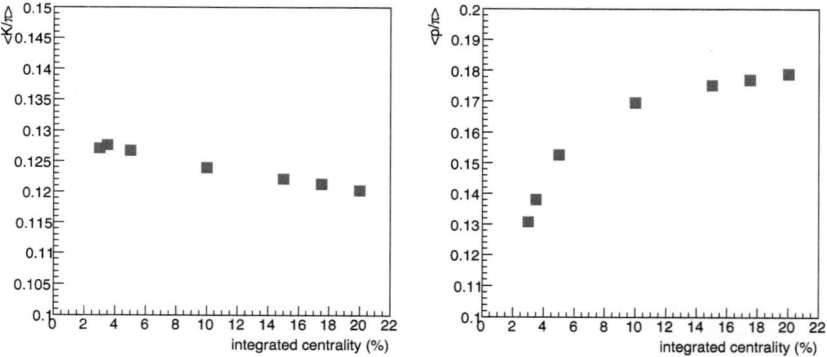

Figure 4.9: Dependence of the mean K/π and p/π ratios on the centrality bin size (data set 01J, see table 3.1). Note the suppressed zero on the Y-axis.

4.2. DEPENDENCE ON CENTRALITY BIN SIZE

The distributions of the event-wise K/π and p/π ratios for data and mixed events are shown in figure 4.10 for selected centrality bin sizes. The plots for all centrality bin sizes can be found in Appendix A. The relative widths of data and mixed events distributions together with the dynamical fluctuations are given in the legends. As expected statistical fluctuations increase with increasing centrality bin size.

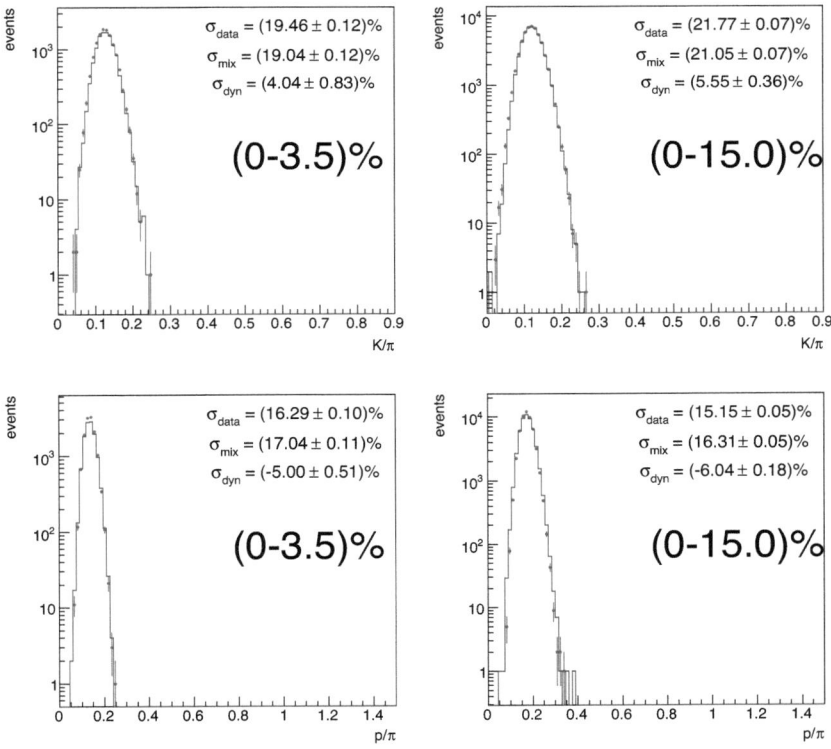

Figure 4.10: Distributions of the event-by-event K/π (upper) and p/π (lower row) ratios for data (points) and mixed (histogram) events for different centrality bin widths as indicated in the figure starting from most central Pb + Pb collisions at 158A GeV (data set 01J, see table 3.1).

The extracted dependence of the relative width of the K/π ratio from data and mixed events on the centrality bin size is shown in figure 4.11. The width increases linearly with increasing bin size due to finite number statistics.

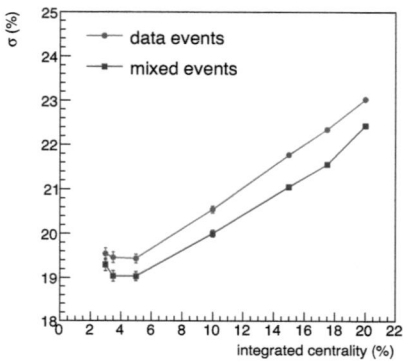

Figure 4.11: Relative width (in %) of data and mixed events as a function of centrality bin width starting from most central Pb + Pb collisions at 158A GeV (data set 01J, see table 3.1).

The dynamical fluctuations of the K/π and p/π ratios as a function of the centrality bin size is shown in figure 4.12.

Comparison of the dependences from two data sets (01J and 01I) are shown in figure 4.13.

A good agreement between results from the minimum bias data set from 2000 (low intensity, 01J) and the central data set from 1996 with a selected centrality bin size of 3.5% is seen. However they both differ by (1.5-2)% from the data set from 2000 (high intensity, 01I). During the data taking in 2000 with high intensity beam (data set 01I) 256 time bins were used in the MTPCs configuration (instead of standard 512 time bins). The comparison of the dE/dx resolution for the data sets 01I and 01J shows that in case of reduced number of time bins the resolution becomes worse by (1.5-2)%. Also all corrections have originally been tuned for 512 time bins, so in general the quality of the dE/dx measurement is expected to be poorer for 256 time bins. Thus the performance of the particle identification by dE/dx measurement decreases, and the systematic error of the dynamical fluctuations of the particle yield ratio increases. This might explain the difference in the results observed in this analysis.

4.2. DEPENDENCE ON CENTRALITY BIN SIZE

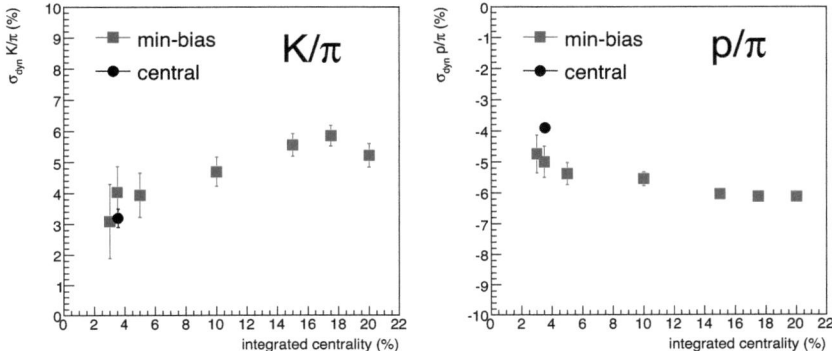

Figure 4.12: Dynamical fluctuations of the K/π and p/π ratios as a function of increasing centrality bin width for semi-central Pb + Pb collisions at 158A GeV.

For both, K/π and p/π ratio fluctuations only little bin size dependence is seen. The difference between the results for a bin width of 3.5% and of 5% is small (in the order of 0.5%). For the study of the centrality dependence of the fluctuation signal centrality bins of 5% and 10% (last for estimation of the systematic errors) were chosen.

4.2.2 Semi-peripheral events

The same study as presented in section 4.2.1 was performed starting from peripheral events (centrality of 30%) in order to check wether the same dependences on bin width are observed as for the most central events. Distributions of the event-by-event K/π and p/π ratios are depicted in figure 4.14 for selected centrality bin sizes. The distributions for all considered bin sizes can be found in Appendix A. The relative widths of data and mixed events distributions together with the dynamical fluctuations are given in the legends. As expected statistical fluctuations increase with increasing centrality bin size.

54 CHAPTER 4. RESULTS OF THE DATA ANALYSIS

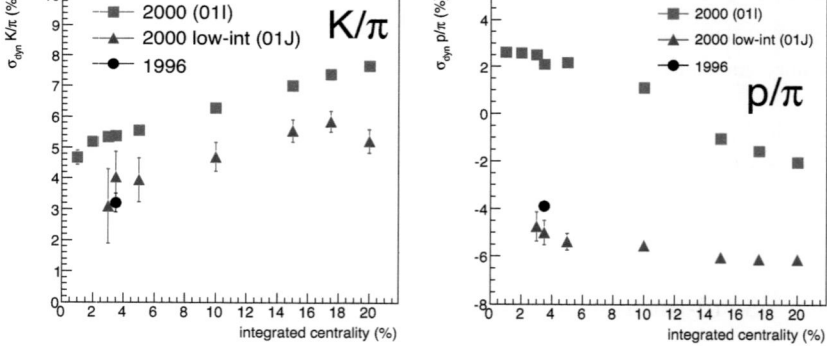

Figure 4.13: Comparison of dynamical fluctuations of the K/π and p/π ratios as a function of increasing centrality bin width for semi-central Pb + Pb collisions at 158A GeV for data sets 01J and 01I.

4.2. DEPENDENCE ON CENTRALITY BIN SIZE

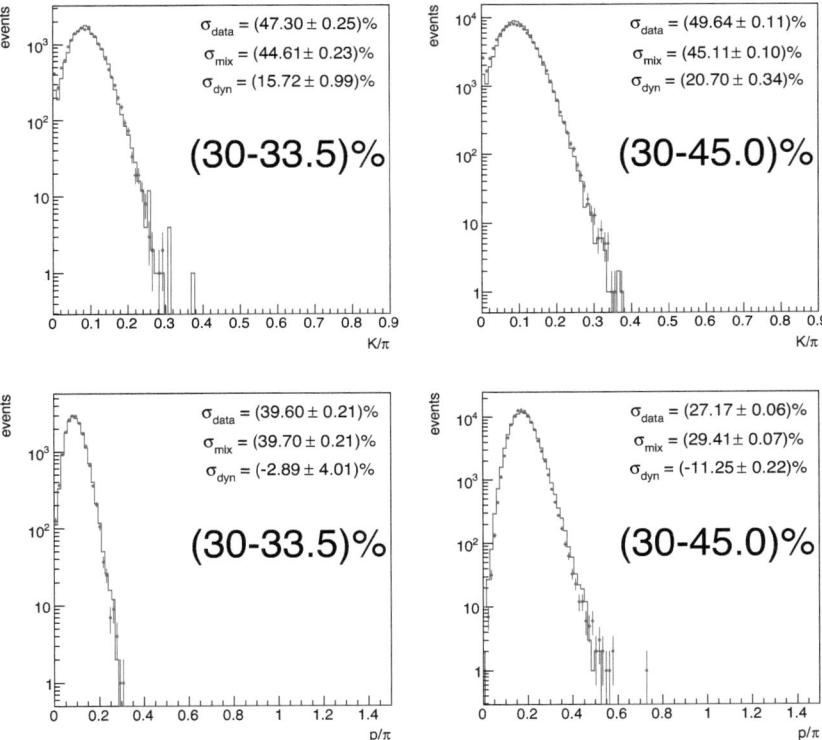

Figure 4.14: Distributions of the event-by-event K/π and p/π ratios for data (points) and mixed events (histogram) for different centrality bin width for semi-peripheral Pb + Pb collisions at $158A$ GeV beam energy (data set 01J).

Dynamical fluctuations of the K/π and p/π ratios as a function of the centrality bin size is shown in figure 4.15.

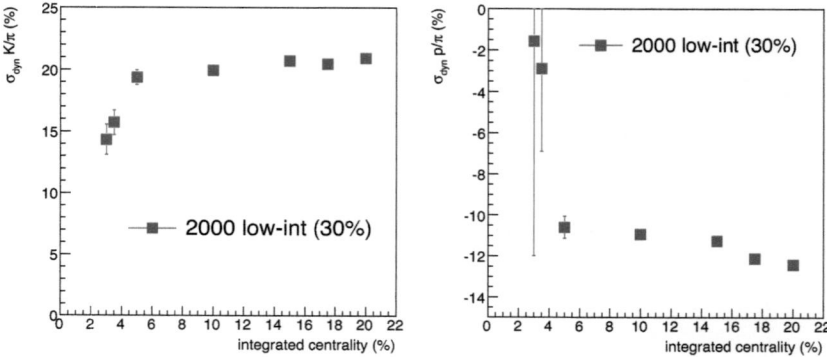

Figure 4.15: Dynamical fluctuations of the K/π and p/π ratios as a function of centyrality bin width for semi-peripheral Pb + Pb collisions.

Observed dependences result in 5% systematic error on the K/π ratio fluctuations and in 2% systematic error on the dynamical fluctuations of the p/π ratio. These systematic errors will be considered in the study of the centrality dependence of the fluctuation signal.

4.3 Centrality dependence

As was shown in the previous subsection, the difference between small centrality bin widths as e.g. 3.5% as used to study the energy dependence and 5% which will be used to study the centrality dependence is in general on a 1 - 2% level and up to 5% for K/π ratio fluctuations in more peripheral events. To evaluate a systematic error a 10% bin size in centrality was also analyzed. We have done systematic studies and used 10% bin size in centrality. Both results will be presented in this subsection.

The centrality of the events was determined using the veto calorimeter of NA49 which detects almost all beam particles and projectile spectators. The distribution of the energy measured by the veto calorimeter in minimum bias Pb + Pb collisions at $158A$ GeV is shown in figure 4.16. Vertical lines show 5% centrality bins from 5 to 55%. This data set contains 59.9% most central collisions.

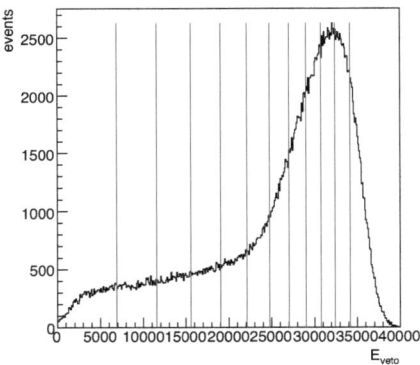

Figure 4.16: Distribution of the veto calorimeter energy for minimum bias Pb + Pb collisions at $158A$ GeV. Vertical lines show 5% centrality bins from 5 to 55%. This data set contains 59.9% most central collisions.

The distributions of the event-by-event K/π and p/π ratios for data and mixed events for selected centrality bins of minimum bias Pb + Pb collisions at $158A$ GeV is shown in figure 4.17. Right shows the ratio of data to mixed events. Plots for all considered centrality bins can be found in Appendix B. The relative widths of data and mixed events distributions together with the dynamical fluctuations are given in the legends. As expected statistical fluctuations increase with decreasing centrality of Pb + Pb collisions.

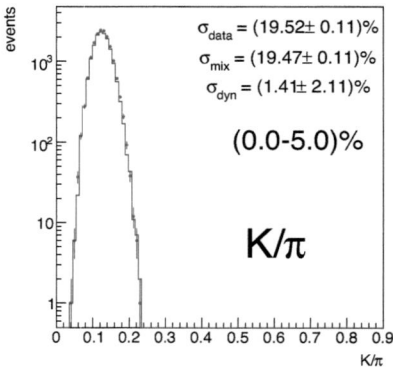
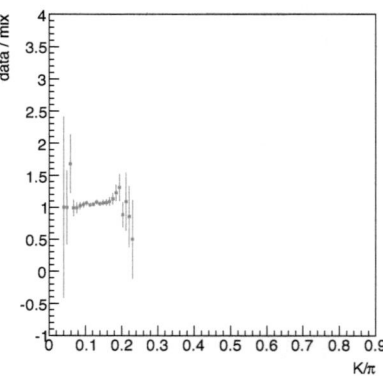
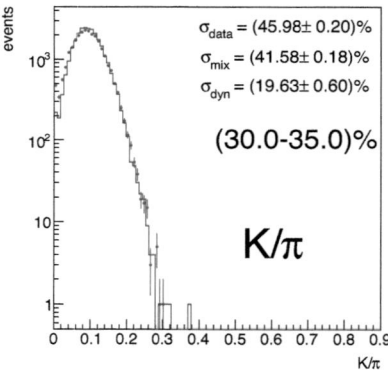
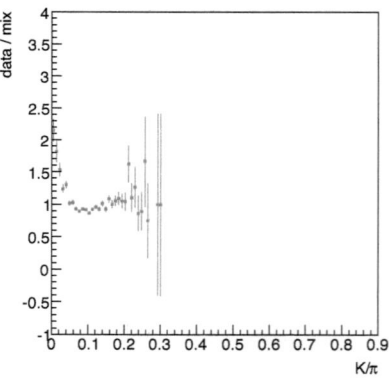

4.3. CENTRALITY DEPENDENCE

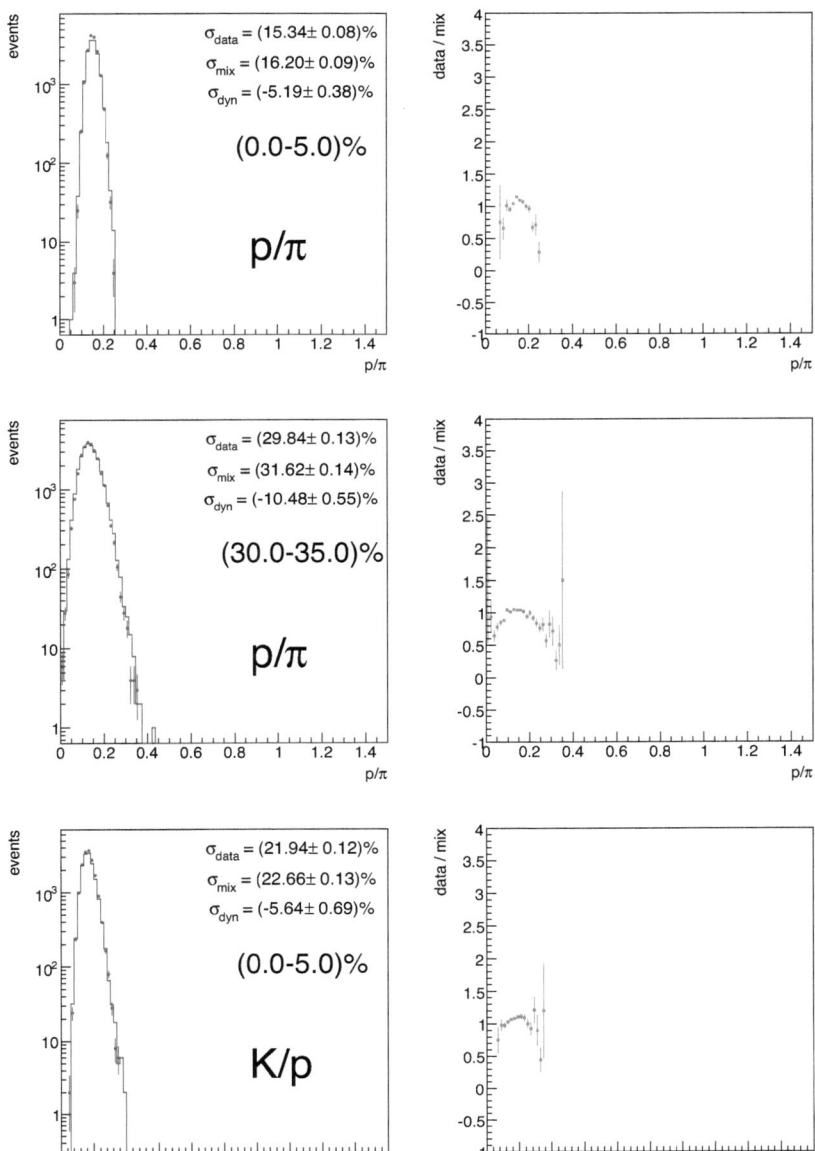

60 CHAPTER 4. RESULTS OF THE DATA ANALYSIS

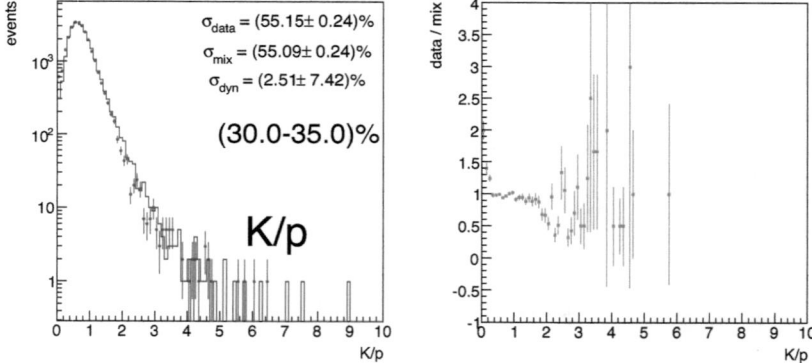

Figure 4.17: Distributions of the event-by-event K/π, p/π and K/p ratios for data and mixed events for selected centrality bins of minimum bias Pb + Pb collisions at $158A$ GeV. Right panels show the ratio of data to mixed events.

4.3. CENTRALITY DEPENDENCE

Around a centrality of 30% the multiplicity of kaons becomes low enough (on average 10 kaons per event) developing a prominent spike at zero, which is also observed in the distribution of the kaon to pion yield ratio at the lowest SPS energy for 3.5% most central events. For peripheral Pb + Pb collisions a large tail in the eventwise K/p ratio distribution is observed, which is not seen in other ratios.

The centrality dependence of the dynamical fluctuations of the K/π, p/π and K/p ratios as the main result of this thesis is shown in figure 4.18.

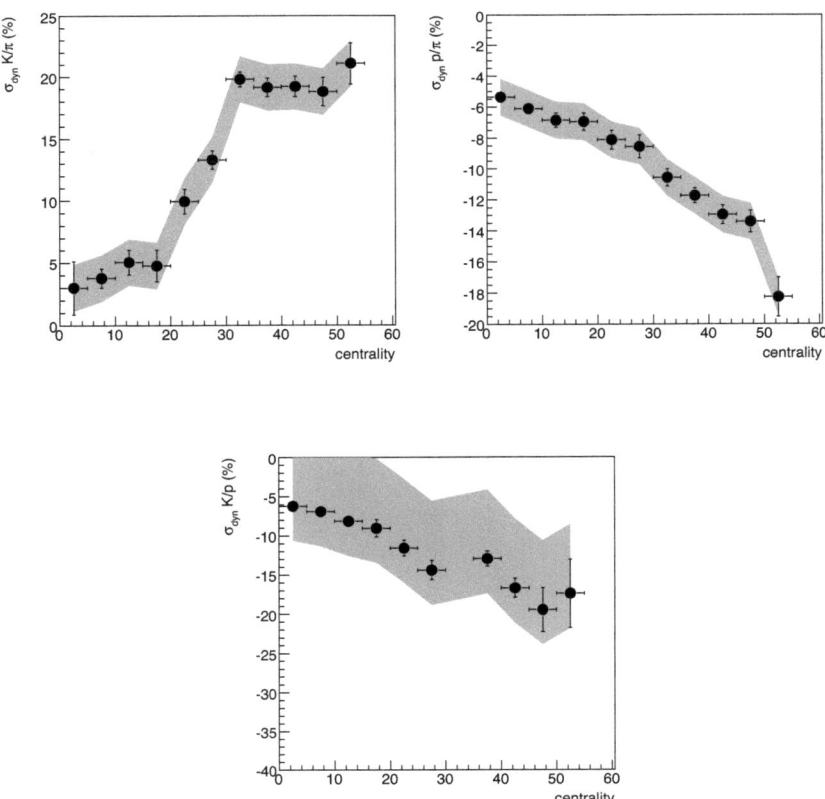

Figure 4.18: Centrality dependence of the dynamical fluctuations of the kaon to pion yield ratio for minimum bias Pb + Pb collisions at 158A GeV.

Chapter 5

Discussion of the results

The Value of the K/π ratio fluctuations increases with decreasing centrality untill a centrality of 35% and then saturates at around 20%. By ratio fluctuations we mean the event-by-event dynamical fluctuations of the particle yield ratio. The point where fluctuations start to saturate coincides with the development of a pronounced spike at zero. One possible explanation of such a saturation is that this artificial increase of entries at zero prevents the data distribution from becomeing broader on the left side in comparison to the distribution of the mixed events, thus the geometrical difference stays constant.

The centrality of the collisions was translated into the number of wounded nucleons per collision using model calculations [44]. The dependence between these two variables is almost linear.

The average extracted yields per event of the π, K and p as a function of number of wounded nucleons are shown in figure 5.1.

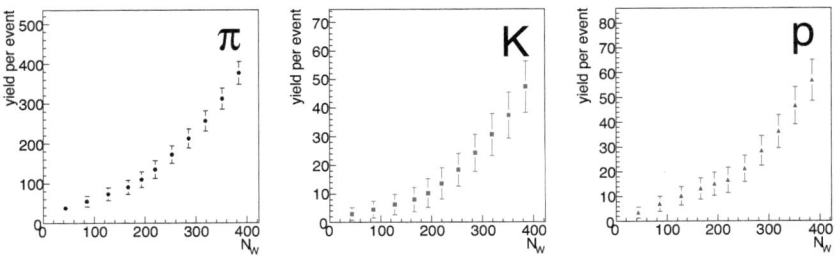

Figure 5.1: Average yields of π, K and p as a function of number of wounded nucleons in Pb + Pb collisions at 158A GeV.

The kinky proton behaiviour can be explained as the variation of the dn/dy shapes with centrality.

Figure 5.2 shows the dependence of the mean kaon to pion, proton to pion and kaon to proton yield ratios on the number of wounded nucleons. The mean ratio goes up with increasing centrality.

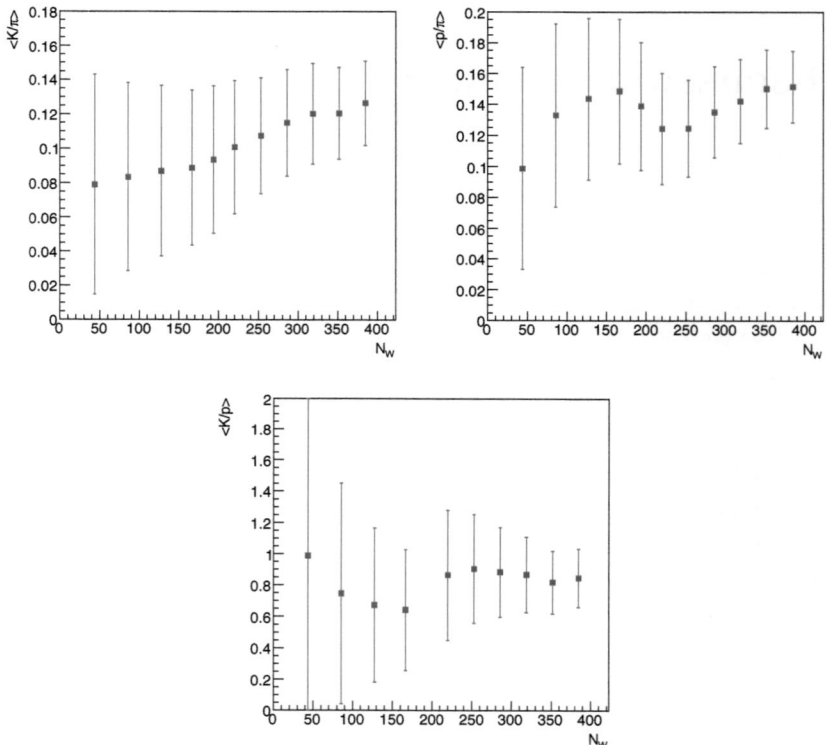

Figure 5.2: Dependence of the mean K/π (left top), p/π (right top) and K/p (bottom) ratios on the number of wounded nucleons. The loose set of track cuts was used here.

The kinky structures in the proton to pion and kaon to proton ratios are due to the strucutres in the proton and kaon yields.

The dependence of the relative width of the eventwise kaon to pion, proton to pion and kaon to proton yield ratios for data events on the number of wounded nucleons is shown in figure 5.3. Due to finite number statistics the relative width is decreasing with increasing centrality of Pb + Pb collisions.

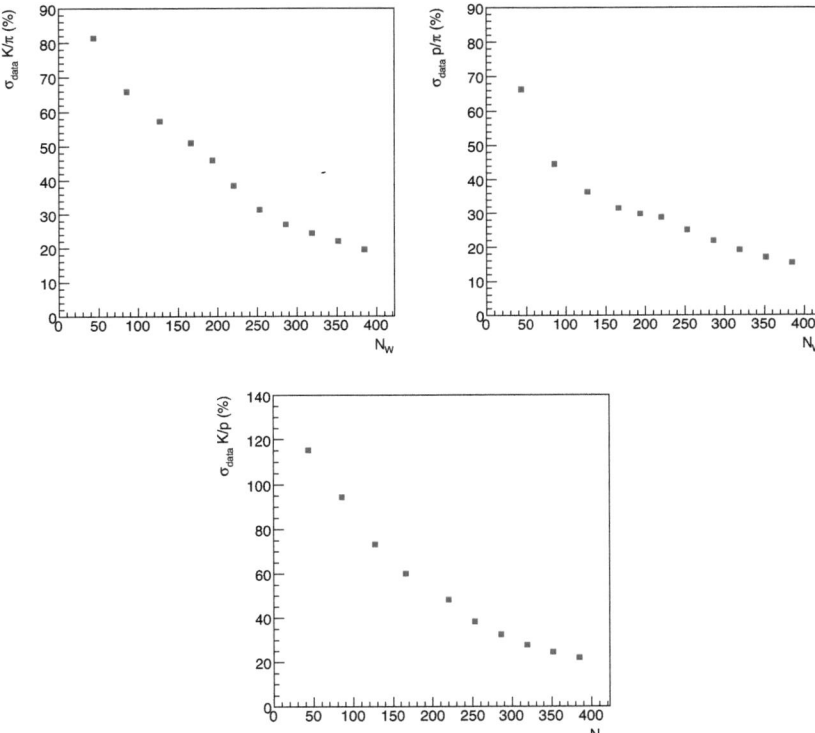

Figure 5.3: Dependence of the relative width of the eventwise kaon to pion, proton to pion and kaon to proton yield ratios distribution on the number of wounded nucleons. The loose set of track cuts was used here.

Figure 5.4 shows the dependence of the dynamical fluctuations of the K/π, p/π and K/p ratios on the number of wounded nucleons in Pb + Pb collision at 158A GeV beam energy. Systematic errors shown here by the band were estimated as the difference between results for the loose and tight track cuts, taking also into account the observed dependence of the dynamical fluctuations on the centrality bin size. The points on the plots represent the mean values of the results for two different set of track cuts and the results for 5% and 10% centrality bin widths. The absolute values of the dynamical fluctuations increase with the decreasing centrality. And approaches zero for central Pb + Pb collisions. Saturation of the K/π and K/p ratio flucutations is observed for peripheral collisions which can be attributed to the development of the spike at zero.

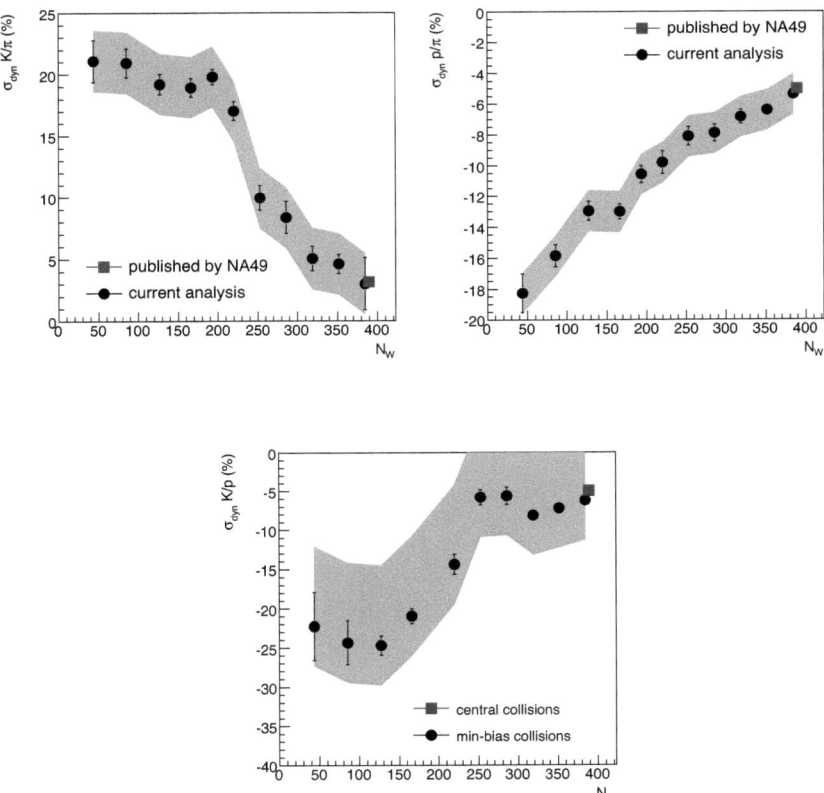

Figure 5.4: Dependence of the dynamical fluctuations of the K/π, p/π and K/p ratios on number of wounded nucleons in minimum bias Pb + Pb collisions at 158A GeV. The band shows the estimated systematic error by varying the track cuts and the centrality bin width. The points are the mean values.

Due to large multiplicity of protons no spike at zero in the distribution of the proton to pion yield ratio even for peripheral collisions is observed. Thus there is no leveling of the dynamical fluctuations of the p/π ratio and they follow the one over number of wounded nucleons scaling in the complete range. Similar to the behaiviour of the dynamical fluctuations of the kaon to pion yield ratio we observe a leveling of the dynamical fluctuations of the K/p ratio but at the number of wounded nucleons of 100. We expect, that this observation is also connected with low kaon multiplicity in peripheral events.

The relative contribution of the spike at zero to the eventwise K/π and K/p distributions is shown in figure 5.5 versus centrality of Pb + Pb collisions at 158A GeV beam energy. The enhanced first bin exceeds the level of 2% at a centrality of 30% for the K/π and of 40% for the K/p ratio distributions.

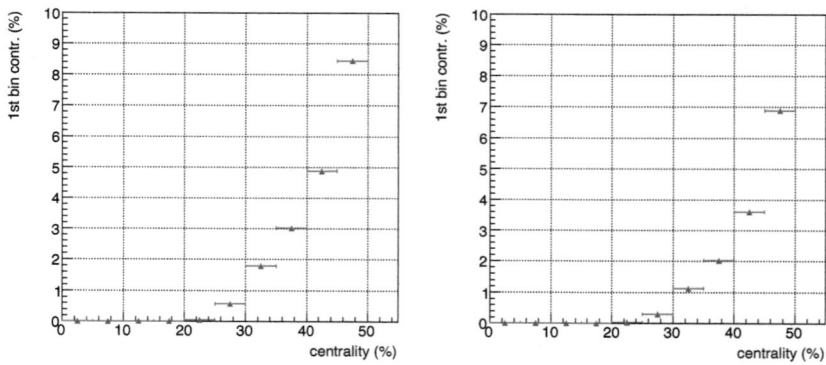

Figure 5.5: The relative contribution of the spike at zero in the eventwise K/π (left picture) and K/p (right picture) distributions for data events versus centrality of Pb + Pb collisions at 158A GeV beam energy.

The centrality dependence of the dynamical fluctuations of the K/π ratio with the suppressed spike at zero is shown in figure 5.6. Instead of saturation a decrease of the fluctuations is observed for peripheral collisions.

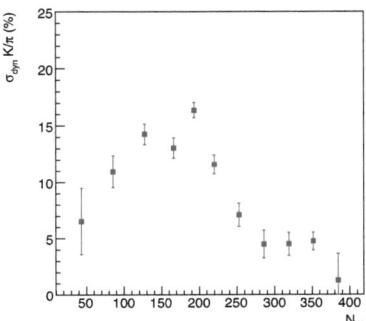

Figure 5.6: Centrality dependence of the dynamical K/π ratio fluctuations with the suppressed spike at zero for Pb + Pb collisions at 158A GeV beam energy.

For the estimation of the systematic errors 10% bin size in centrality was used. Comparison of the dynamical fluctuations of the K/π ratio for 5% and 10 bin size in centrality of Pb + Pb collisions at 158A GeV is shown in figure 5.7.

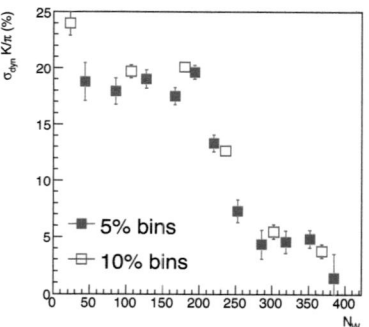

Figure 5.7: Centrality dependence of the dynamical K/π ratio fluctuations for 5% and 10% bin sizes in centrality of Pb + Pb collisions at 158A GeV beam energy.

5.1 UrQMD simulations

The results in this section have been obtained with UrQMD version 1.3 [37] by analyzing the freeze-out configuration. Consequently, weak decays are not included in the analysis since they are not implemented in the UrQMD model. This is justified by the fact that the CBM experiment is able to exclude secondary particles efficiently from the analysis by a cut on the track impact parameter at the event vertex. However this might be problematic for the NA49 experiment, which for example detects 50% of pions from K^0 decay as primary pions.

5.1.1 Influence of detector acceptance

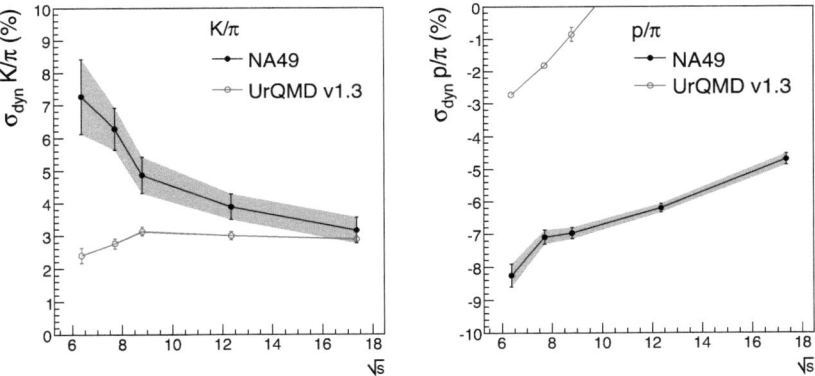

Figure 5.8: Dynamical event-by-event fluctuations of the kaon to pion (left) and proton to pion (right) ratios as function of $\sqrt{s}$. The open circles show values obtained by UrQMD simulations in 4π, the closed ones NA49 data [45].

After the subtraction of the non dynamical background, a limited detector acceptance may still affect the measured fluctuations. In the context of resonance decays, the acceptance influences the mean multiplicities of independently produced particles serving as normalisation (see sections 5.2.1 and 5.2.2) and may destroy correlations of decay products.

In order to approximate the NA49 acceptance, we restrict our analysis to tracks in the forward hemisphere and to momenta larger than 3 GeV [45]. In addition, particles near

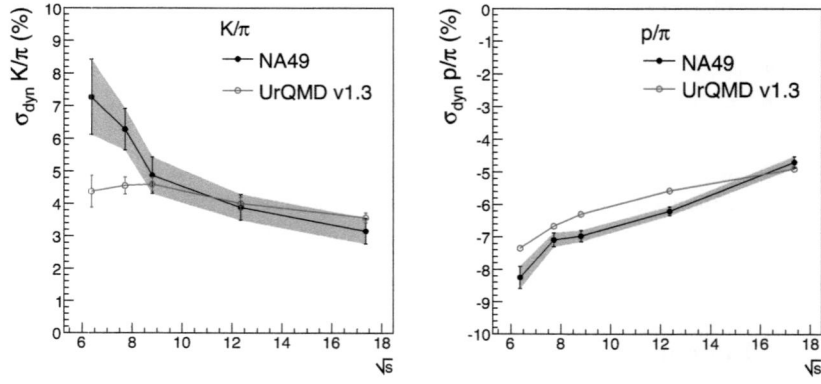

Figure 5.9: Dynamical event-by-event fluctuations of the kaon to pion (left) and proton to pion (right) ratios as function of $\sqrt{s}$. The open circles show values obtained by UrQMD simulations using approximated NA49 acceptance cuts, the closed symbols show NA49 data [45].

beam rapidity are cut to suppress projectile spectators (rapidity should be smaller than beam rapidity - 0.5). Results for the kaon to pion fluctuations in 4π in the whole energy range measured by NA49 are shown in Figure 5.8, those after the acceptance cuts in Figure 5.9, both together with NA49 data [45]. The comparison of the two figures shows that the acceptance has little effect on the kaon to pion ratio; in particular it does not introduce a significant energy dependence. In contrast, the acceptance strongly influences the proton to pion ratio fluctuations, which can be attributed to the fact that the proton spectators are cut out if restricting the analysis to the NA49 acceptance. The numerical values of the fluctuations of various particle ratios in 4π and after NA49 acceptance cuts are compared in Table 5.1 for central Au+Au collisions at $25A$ GeV beam energy as relevant for the future experiment CBM at FAIR.

The UrQMD model results agree reasonably well with the earlier UrQMD calculations [45], showing that the analysis algorithms are consistent as the same UrQMD version was used. Small remaining differences can be explained by the use of only approximated acceptance cuts in this calculation neglecting the incomplete azimuthal acceptance of NA49.

5.1. URQMD SIMULATIONS

Particle ratio	Dynamical fluctuations (%)	
	4π	after appr. NA49 acceptance cuts
$(K^++K^-)/(\pi^++\pi^-)$	2.5 ± 0.3	3.4 ± 0.6
$(p+\bar{p})/(\pi^++\pi^-)$	-5.56 ± 0.04	-7.0 ± 0.1
K^+/π^+	-6.1 ± 0.1	-6.1 ± 0.5
K^+/π^-	-8.0 ± 0.1	-8.6 ± 0.3
K^-/π^+	-8.7 ± 0.3	-10.7 ± 0.7
K^-/π^-	-7.3 ± 0.3	-6.7 ± 1.2

Table 5.1: Dynamical fluctuations of various particle ratios in 4π and within an approximated NA49 acceptance. Simulations with UrQMD for central (b=0 fm) Au + Au collisions at 25A GeV beam energy.

5.1.2 dE/dx resolution

The transport model UrQMD provides kinematics and particle ID of all particles created in heavy ion collisions. So, within this model one has a reference and can study effects such like the influence of detector resolution on dynamical fluctuations. Since the spike at zero in the eventwise K/π ratio is most pronounced at the lowest SPS energies, central Pb + Pb collisions at 20A GeV beam energy were simulated using the UrQMD model version 1.3 [35]. Accepted tracks were selected according to the detailed NA49 acceptance table for this beam momentum and for the set of loose track cuts [32]. The acceptance is shown in figure 5.10.

For each accepted track a dE/dx value was simulated using the Probability Density Functions (PDFs) from the inclusive fit of the data events at 20A GeV. The PDFs were tabulated in total momentum, transverse momentum and azimuthal angle (p_{tot}, p_t and ϕ) and contain the mean dE/dx value and the dE/dx resolution for each of these bins. These simulated dE/dx values were processed through the same analysis procedure as tracks from data, i.e. an inclusive fit for obtaining the PDFs (for UrQMD) and the event-by-event fit for getting the kaon to pion ratio in each event was performed. For reference, kaons and pions were simply counted using the Monte Carlo (MC) information. Figure 5.11 shows the distributions of the eventwise kaon to pion ratio for the MC identification and the event-by-event fit.

As expected, the distribution becomes broader when going from MC identification to the event-by-event fit including the dE/dx resolution. The interesting observation is that in case of the event-by-event fit the spike at zero appears, as it was seen for data events.

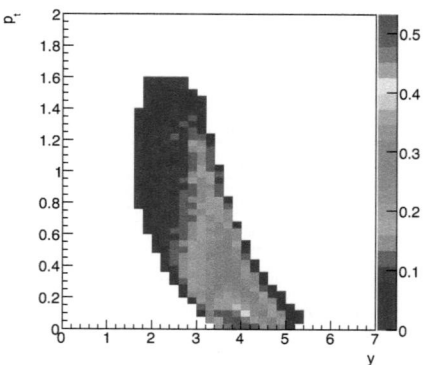

Figure 5.10: Acceptance table for the NA49 detector. Calculated for 20A GeV and integrated over ϕ bins.

Obviously, this is a feature appearing if kaon numbers are too small. One could avoid this spike at zero by allowing negative kaon yileds in the fit. This was done as a check but resulted in fit instabilities. The fit becomes unstable because the minimum of the likelihood can not be found when allowing for negative yields.

After the event mixing procedure, which destroys all correlations inside the real events and provides a statistical reference, dynamical fluctuations of the kaon to pion ratio were calculated for UrQMD events for both methods. Figure 5.12 shows the distributions of the event-by-event kaon to pion ratio for data and mixed events both for MC identification (left plot) and the event-by-event fit (right plot.)

The dynamical fluctuations are the same within errors for both methods, so dE/dx resolution does not bias the result although the spike at zero appears.

5.1.3 Centrality dependence

The effect of dE/dx resolution on the dynamical fluctuations of the K/π ratio has been studied for peripheral Pb + Pb collisions were the spike at zero in the eventwise ratio distribution is much larger. For this analysis minimum bias Pb + Pb collisions at 158A GeV have been simulated using the UrQMD model [35]. Simulated events were selected in 5% centrality bins using the impact parameter of the collisions. The distribution of the impact parameters for all events is shown in figure 5.13. Vertical lines show the applied

5.1. URQMD SIMULATIONS

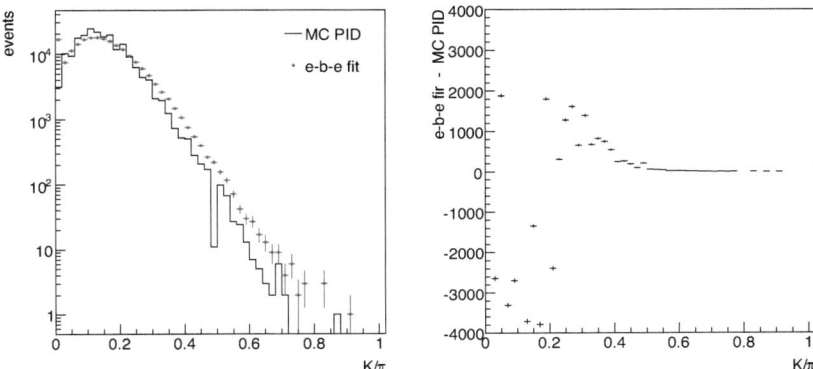

Figure 5.11: Distribution of the eventwise kaon to pion ratio for MC identification and event-by-event fit (left plot) and the difference between them (right plot).

cuts.

In this study we have assumed perfect particle identification, provided by UrQMD, as well as the event-by-event fit using the simulated dE/dx values, as was described in the previous section. In order to estimate the background, the same events were mixed only within the same centrality class. The dynamical fluctuations of the kaon to pion yield ratio in the 4π acceptance as a function of centrality of the event is shown in figure 5.14.

In addition a possible acceptance effect has been simulated by applying the NA49 acceptance filter [32]. Centrality dependence of the dynamical K/π ratio fluctuations within the acceptance is shown in figure 5.15.

Both particle identifications methods start to differ only for the very peripheral Pb + Pb collisions. Approximately at the same centrality the contribution of the spike at zero in the eventwise yield ratio distribution becomes more than 2%. So, this deviation from the MC truth can be attributed to the artifact of the event-by-event fit in case of low kaon multiplicity. For $20A$ GeV beam energy the contribution of the spike at zero is in the order of 3%, thus if there is a bias in the results for the lowest SPS energy, than the dynamical fluctuations of the kaon to pion yield ratio are lowered.

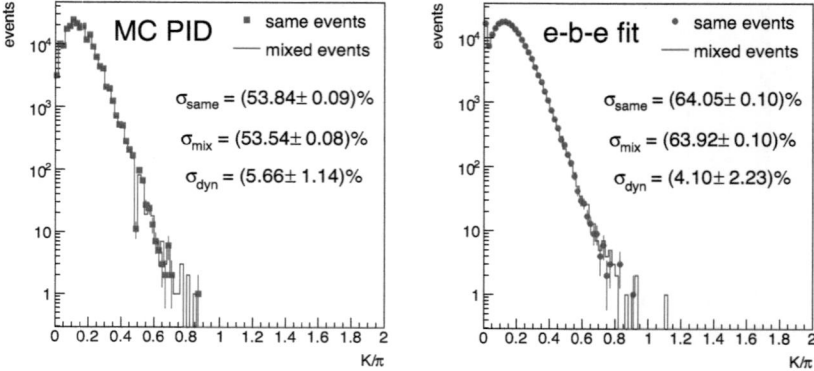

Figure 5.12: Distributions of the eventwise kaon to pion ratio for same and mixed events both for MC identification (left plot) and the event-by-event fit (right plot).

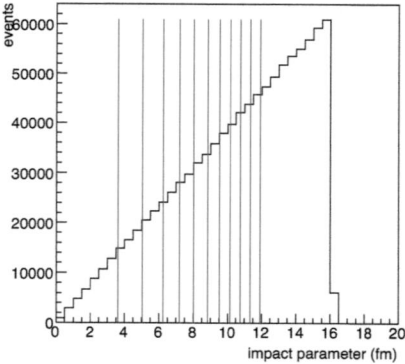

Figure 5.13: Impact parameter distribution for minimum bias Pb + Pb collisions at $158A$ GeV simulated with the UrQMD model. With vertical lines a centrality selection in 5% bins is shown.

5.1. URQMD SIMULATIONS

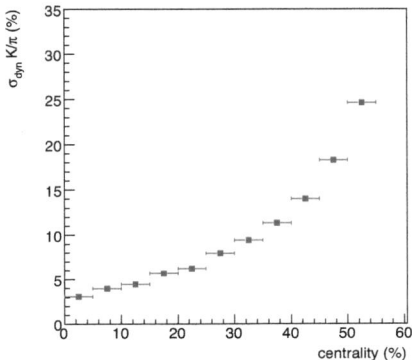

Figure 5.14: Dynamical fluctuations of the kaon to pion yield ratio in 4π as a function of the centrality of Pb + Pb collisions at $158A$ GeV (simulations with UrQMD).

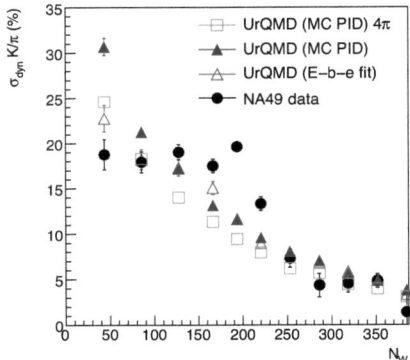

Figure 5.15: Dynamical fluctuations of the kaon to pion yield ratio within the acceptance of NA49 as a function of the centrality of Pb + Pb collisions at $158A$ GeV. The results for the MC identification and the event-by-event fit are shown together with the NA49 data and values in the 4π acceptance.

5.1.4 Resonance contributions in UrQMD

The K/π fluctuations measured by NA49 refer to the sum of charged kaons over the sum of charged pions. As resonances are expected to feed all channels, e.g. prominently the $p\pi$ channel but also $K\pi$, e.g. via K^* decays, we studied the single particle ratios K^+/π^+, K^+/π^- and vice-versa. Figure 5.16 shows the distributions of these ratios in central (b=0 fm) Au + Au collisions at $25A$ GeV, both for same and mixed events together with the numerical value of the dynamical fluctuations. Remarkably, all of these fluctuations are negative, signalling a correlation due to resonance decays. This holds also for the ratios K^+/π^+ and K^-/π^- which are fed by the decay channels of the higher lying K_1 resonance $K_1 \to K\rho \to K\pi\pi$ and $K_1 \to K^*\pi \to K\pi\pi$.

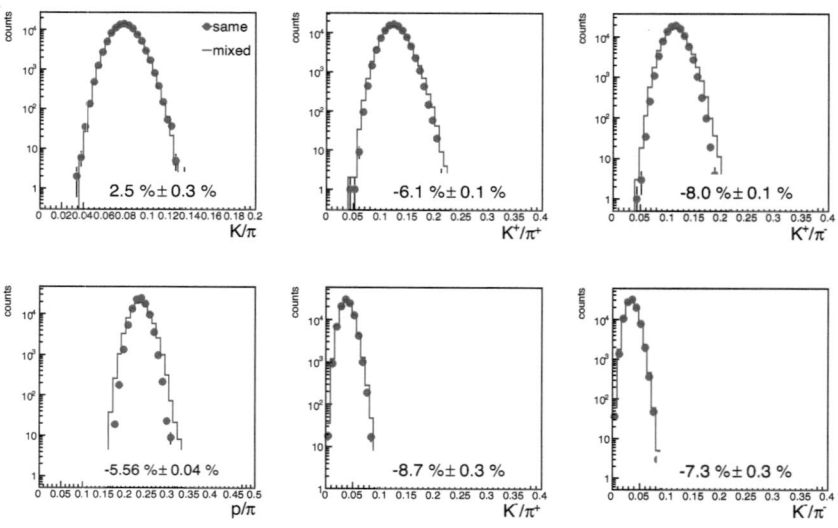

Figure 5.16: Distributions of the event-wise particle ratios for same and mixed events together with the values of dynamical fluctuations. Simulations done with UrQMD for central (b=0 fm) Au + Au collisions at $25A$ GeV beam energy. Left column: sum of kaons over sum of pions (top), sum of protons and anti protons over sum of pions (bottom). Middle and right columns: single particle ratios.

To understand the relative importance of resonance feeddown for dynamical fluctuations, the sources of kaons in UrQMD have been studied using the full collision history.

5.1. URQMD SIMULATIONS

Figure 5.17 shows the contributions of various resonances to the final state kaons. About half of the K^+ and one third of K^- originate from the decay of K^*, while the K_1 is less often produced and thererfore contributes less to the kaon yields. We thus can qualitatively understand the source of the negative fluctuations in the single particle K/π ratios seen in UrQMD.

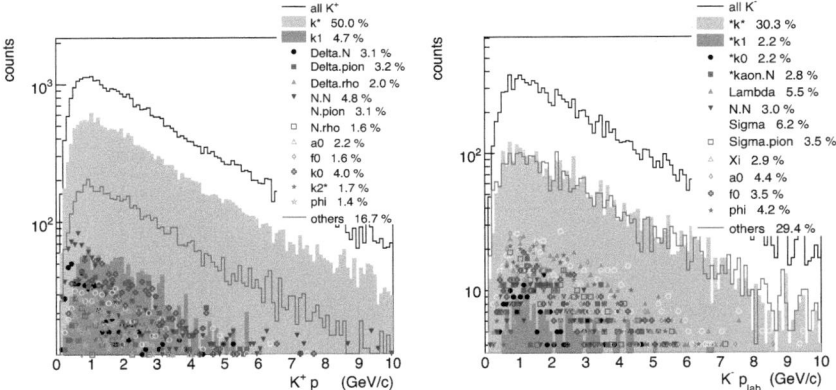

Figure 5.17: Momentum distribution of K^+ (left) and K^- (right) generated by UrQMD for central (b=0 fm) Au + Au collisions at 25A GeV beam energy. Black lines show the momentum spectra of all $K^{+/-}$, the other distributions (momentum spectra of kaons from different sources) are named in the legends. The results were obtained by analysing the UrQMD collision history file.

The positive fluctuation value in the K/π ratio including all charges is thus puzzling at first sight. A possible reason can be ϕ meson decays, which proceed alternatively into a kaon pair or into three pions, either directly or via the ρ-meson. This introduces an anticorrelation for kaons and pions which would be reflected in positive fluctuations. However, according to UrQMD the ϕ meson contributes only 1.5 % to the K^+ and 4.3 % to the K^- yield. This puzzle will be resolved in the following section.

In order to check these qualitative considerations, UrQMD calculations were performed suppressing all excited K and ϕ decays. The simulations are again done for central (b=0 fm) Au + Au collisions at 25A GeV beam energy. Table 5.2 summarises the results obtained in 4π. As expected, the proton to pion ratio fluctuations are not affected at all by the supression of the strange resonance decays. As can be seen from the numbers, the excited kaons indeed change the K/π ratios towards smaller numbers. However the effect is small. Only the K/π ratio fluctuations including all charges are stronger reduced. The suppression of the ϕ decay leaves the single charged particle ratios untouched while it slightly reduces the all charged K/π fluctuations. Again, this agrees qualitatively with the expectations. However, other sources of correlations are obviously present in particular for the single particle ratios.

Particle ratio	Dynamical fluctuations (%)		
	Standard	no K^*	no ϕ
$(K^++K^-)/(\pi^++\pi^-)$	2.5 ± 0.3	6.7 ± 0.2	1.3 ± 0.5
$(p+\bar{p})/(\pi^++\pi^-)$	-5.56 ± 0.04	-5.59 ± 0.04	-5.51 ± 0.04
K^+/π^+	-6.1 ± 0.1	-3.6 ± 0.6	-6.3 ± 0.1
K^+/π^-	-8.0 ± 0.1	-6.3 ± 0.3	-8.2 ± 0.1
K^-/π^+	-8.7 ± 0.3	-8.1 ± 0.3	-8.8 ± 0.3
K^-/π^-	-7.3 ± 0.3	-6.6 ± 0.4	-7.3 ± 0.4

Table 5.2: Dynamical fluctuations of various particle ratios calculated with UrQMD. Values are shown in the 4π acceptance including all resonances ("Standard"), and alternatively with a suppression of K^* decays or ϕ-meson decays.

It has to be noted that the suppression of the K^* decay significantly influences the mean kaon yields, thus making other correlations more visible (see sections 5.2.1 and 5.2.2). It is hence difficult to get a more quantitative insight into the resonance decay dynamics by using UrQMD in this manner.

5.2 Simulations with phase space models

In order to study the influence of resonance decays on the dynamical particle ratio fluctuations, we generate events containing kaons, pions and resonances with given mean multiplicities and phase space distributions. For this study we restrict ourselves to the 4π acceptance. Resonance decays are performed for predefined decay channels until stable daughters are reached. Thus, their influence on the fluctuations can be studied in an environment of a fixed number of independently produced particles.

The simulations were performed with mean multiplicites of pions and kaons taken from UrQMD calculations for central Au + Au collisions at 25 AGeV beam energy. The resonance yields were treated as free parameters.

5.2. SIMULATIONS WITH PHASE SPACE MODELS

5.2.1 Influence of the $K^*(892)$ decay

For a given mean number of independently produced pions and kaons, the fluctuations induced by the $K^* \to K^+\pi^-$ decay can be obtained analytically (see also [46]). The generic analytical exprssion for the σ_{dyn} is given by:

$$\sigma_{dyn} = \sqrt{\frac{var(N_K)}{<N_K>^2} + \frac{var(N_\pi)}{<N_\pi>^2} - 2\frac{cov(N_K, N_\pi)}{<N_K><N_\pi>}}, \quad (5.1)$$

Assuming a Poissonian distribution of N_{π^-} and N_{K^+}, the relative width of the event-wise K^+/π^- ratio distribution for same events consists of three terms :

$$\sigma^2_{same} = \frac{1}{<N_{K^+}>} + \frac{1}{<N_{\pi^-}>} - 2\frac{Cov(N_{K^+}, N_{\pi^-})}{<N_{K^+}><N_{\pi^-}>} \quad (5.2)$$

In the mixed events all correlations are destroyed, and the third terms vanishes. Note that by construction, the mean multiplcities remain unchanged:

$$\sigma^2_{mixed} = \frac{1}{<N_{K^+}>} + \frac{1}{<N_{\pi^-}>} \quad (5.3)$$

So we have $\sigma_{same} < \sigma_{mixed}$ (correlation) and according to (3.9), the dynamical fluctuations are

$$\sigma_{dyn} = -\sqrt{\sigma^2_{mixed} - \sigma^2_{same}} = -\sqrt{2\frac{Cov(N_{K^+}, N_{\pi^-})}{<N_{K^+}><N_{\pi^-}>}} \quad (5.4)$$

The number of kaons consists of independently produced particles ($N^{(p)}_{K^+}$) as well as products from resonance decays ($<K^*> \cdot BR$ on average, where $BR = 50\%$). The same holds for pions. After calculating the covariance we get

$$\sigma_{dyn} = -\sqrt{\frac{2\cdot <K^*> \cdot BR}{(<N^{(p)}_{K^+}> + <K^*> \cdot BR)(<N^{(p)}_{\pi^-}> + <K^*> \cdot BR)}} \quad (5.5)$$

Figure 5.18 shows the result of this generic simulation together with the analytical formula 5.5. A good agreement is observed. The absolute value of the fluctuations induced by the K^* decay increases approximatly proportional to the square root of the relative amount of K^*.

5.2.2 Influence of the $\phi(1020)$ decay

With the same technique, the influence of the ϕ meson on the $(K^+ + K^-)/(\pi^+ + \pi^-)$ ratio fluctuations was studied. The simulated decay scheme of the resonance is shown in Figure 5.19, where only strong decays are considered.

82   CHAPTER 5. DISCUSSION OF THE RESULTS

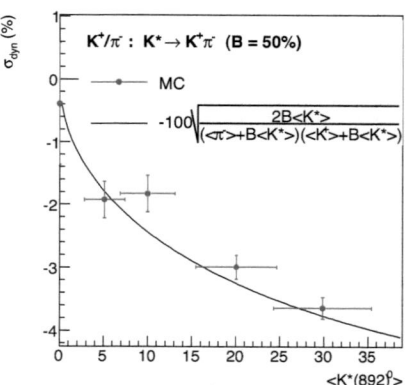

Figure 5.18: Dynamical fluctuations of the kaon to pion ratio calculated with a generic model as function of the $K^*(892)$ yield. Multiplicities of primary pions and kaons are taken from UrQMD calculations for central Au + Au collisions at 25 AGeV beam energy.

Again, an analytical expression can be derived for the fluctuations induced by the ϕ decay. Because of the correlations between K^+ and K^- and between π^+ and π^-, the first two terms of the relative width of the same-event distribution must be expressed in a more general way [46]:

$$\sigma^2_{same} = \frac{Var(N_{K^+}) + Var(N_{K^-}) + 2Cov(N_{K^+}, N_{K^-})}{<N_K>^2} + \frac{Var(N_{\pi^+}) + Var(N_{\pi^-}) + 2Cov(N_{\pi^+}, N_{\pi^-})}{<N_\pi>^2} - 2\frac{Cov(N_K, N_\pi)}{<N_K><N_\pi>}, \quad (5.6)$$

Since N_{K^+} and N_{K^-} are distributed Poissonian like, $Var(N) = N$ for all particles involved, so that the width can be written as

$$\sigma^2_{same} = \frac{1}{<N_K>} + 2\frac{Cov(N_{K^+}, N_{K^-})}{<N_K>^2} + \frac{1}{<N_\pi>} + 2\frac{Cov(N_{\pi^+}, N_{\pi^-})}{<N_\pi>^2} - 2\frac{Cov(N_K, N_\pi)}{<N_K><N_\pi>}, \quad (5.7)$$

For the mixed events the width is driven by statistics only:

$$\sigma^2_{mixed} = \frac{1}{<N_K>} + \frac{1}{<N_\pi>} \quad (5.8)$$

5.2. SIMULATIONS WITH PHASE SPACE MODELS

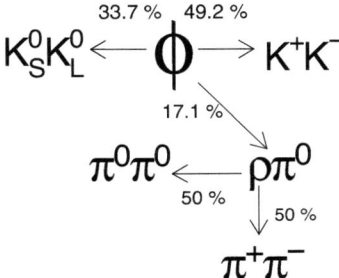

Figure 5.19: The three most probable decay modes of the $\phi(1020)$ meson

Thus we have $Cov(N_K, N_\pi) < 0$ and $\sigma_{same} > \sigma_{mixed}$. Now we can derive the dynamical fluctuations:

$$\sigma_{dyn} = \sqrt{\sigma_{same}^2 - \sigma_{mixed}^2} =$$
$$\sqrt{2\frac{Cov(N_{K^+}, N_{K^-})}{<N_K>^2} + 2\frac{Cov(N_{\pi^+}, N_{\pi^-})}{<N_\pi>^2} - 2\frac{Cov(N_K, N_\pi)}{<N_K><N_\pi>}} \quad (5.9)$$

Analogously to the case with K^* we calculate the covariances and simplify the last equation:

$$\sigma_{dyn} = \sqrt{\frac{2 \cdot <\phi> \cdot BR_{K^+K^-}}{<N_K>^2} + \frac{2 \cdot <\phi> \cdot BR_{\pi^+\pi^-}}{<N_\pi>^2} - 2\frac{Cov(N_K, N_\pi)}{<N_K><N_\pi>}} \quad (5.10)$$

In 4π acceptance we have $N_\pi \gg N_K$ and for the ϕ resonance $BR_{\pi^+\pi^-} \ll BR_{K^+K^-}$, thus the two last terms can be neglected:

$$\sigma_{dyn} \approx \sqrt{\frac{2 \cdot <\phi> \cdot BR_{K^+K^-}}{(<N_K^{(p)}> + 2 \cdot <\phi> \cdot BR_{K^+K^-})^2}} \quad (5.11)$$

Again, the simulation results obtained by the generic resonance model agree well with the derived formula as shown in Figure 5.20, supporting the validity of the approximations performed in the derivation of 5.11. As expected, the ϕ decay results in positive fluctuations of the K/π ratio. The fluctuations increase approximatly proportional to the square root of the ϕ/K ratio. Inspection of eq. 5.9 and 5.11 shows that these fluctuations are not primarily due to the anticorrelation of pion with kaon pairs but due to the covariance of K^+ and K^-.

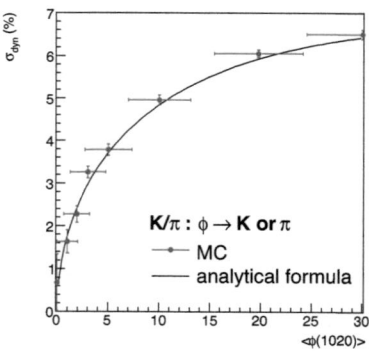

Figure 5.20: Dynamical fluctuations of the kaon to pion ratio obtained by a generic model as function of the $\phi(1020)$ yield. Multiplicities of primary pions and kaons are taken from UrQMD calculations for central Au + Au collisions at 25 AGeV.

It should be noted that for small ϕ multiplicities, as is the case for central Pb + Pb collisions at $20A$ - $30A$ GeV beam energy, the magnitude of the dynamical fluctuations induced by the ϕ decay is a steep function of the ϕ multiplicity, however is small in total value. Transport models should therefore be checked whether they reproduce the measured ϕ yield. This holds in principle for all resonances. However, for the range of ϕ multiplicities allowed by the experimental uncertainties, the fluctuations obtained by our model are well below the K/π ratio fluctuations measured by NA49.

5.3 Scaling of the dynamical fluctuations

In order to understand the nature of the dynamical fluctuations of the particle yield ratios one would like to investigate dependence of those on the number of particles per event in the acceptance. As will be discussed in this section argument of the dependence is different for different particle ratios.

The dynamical fluctuations of the proton to pion yield ratio can be approximated by the correlation term only from the following equation:

$$\sigma_{dyn}^{p/\pi} \approx -\sqrt{\frac{cov(N_p, N_\pi)}{<N_p><N_\pi>}} = -\sqrt{\frac{(<N_p><N_\pi>)^\alpha}{<N_p><N_\pi>}} \quad (5.12)$$

Here the covariance was assumed to be the product of the mean particle multiplicities to the power of α. In case of resonance contributions this parameter α has to be equal to

5.3. SCALING OF THE DYNAMICAL FLUCTUATIONS

0.5 (geometrical mean of the particle yields in the acceptance). Figure 5.21 illustrates the dependence of the event-by-event proton to pion ratio fluctuations on the product of the mean particle multiplicities, comparing the measured energy and centrality dependence of NA49.

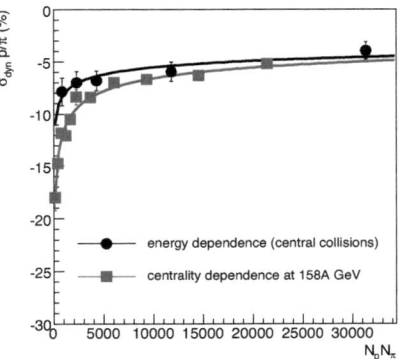

Figure 5.21: Dynamical fluctuations of the proton to pion yield ratio as a function of the product of the mean proton and pion multiplicities.

The data were fitted with equation 5.12 and parameters α were extracted for both energy and centrality dependence of the proton to pion yield ratio fluctuations: $\alpha = 0.66 \pm 0.12$ for the energy dependence and $\alpha = 0.51 \pm 0.03$ for the centrality dependence. The fact that both dependencies follow a common scaling as defined in equation 5.12 strongly supports that the dynamical fluctuations of the proton to pion yield ratio originate from the correlation induced by the baryonic resonance decay into protons and pions.

Since $N_K \ll N_\pi$, the dominating term in the equation for the dynamical fluctuations of the K/π ratio is the kaon term:

$$\sigma_{dyn}^{K/\pi} \approx \sqrt{\frac{var(N_K)}{<N_K>^2}} \quad (5.13)$$

So, one can expect that the dynamical fluctuations of the kaon to pion yield ratio scale with $1/N_K$. The dependence of the fluctuations signal on the number of kaons in the acceptance is shown in figure 5.22 again for both, the energy and centrality dependence.

In the region of low kaon multiplicities, the results for the energy and centrality dependencies are clearly different and they do not follow a common scaling. A possible

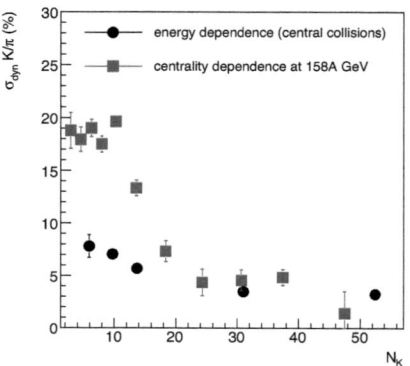

Figure 5.22: Dynamical fluctuations of the kaon to pion yield ratio as a function of the mean kaon multiplicity.

explanation can be the fact that in peripheral Pb + Pb collisions kaons are created in smaller volume, and since strangeness production is correlated with the volume, enhancement of fluctuations for this events can be expected. Calculations with percolation model will be performed in future.

It is difficult to define the leading term for the K/p ratio fluctuations, since $N_K \approx N_p$. In any case, the energy and centrality dependence of the dynamical K/p ratio fluctuations do not scale in the same way. More studies are needed in this direction.

In order to compare the measured energy and centrality dependences of the K/π ratio fluctuations with results obtained by the STAR collaboration these fluctuations were plotted versus midrapidity yields of charged particles (see figure 5.23).

Both dependences are stipper than the results measured by STAR, but they agree with each other within the errors.

5.4 Conclusion

In this work measured by NA49 experiment energy and centrality dependences of the event-by-event dynamical fluctuations of particle yield ratios were presented.

The energy dependence of the K/π ratio fluctuations shows increase towards lower beam energies, which is not reproduced by the transport model UrQMD. The behaviour of the p/π ratio fluctuations is in a good agreement with the results obtained from the transport

5.4. CONCLUSION

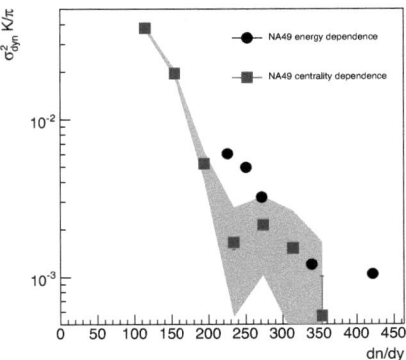

Figure 5.23: The measured energy and centrality dependences as function of midrapidity yields of charged particles.

code. The measured energy dependence of the K/p ratio fluctuations has change of sign and differs from the results of the Monte Carlo simulations.

Centrality dependence was studied with bins in centrality of Pb + Pb collisions with width of 5% and 10% for estimation of the systematic errors. Additional systematic errors were obtained by varying the set of track cuts as was already explained. The measured centrality dependence of particle ratio fluctuations at $158A$ GeV beam energy indicate an increase of the absolute value of the dynamical fluctuations with decreasing centrality for all considered particle yield ratios. A saturation of the dynamical fluctuations of the K/π and K/p ratios was observed for centralities below 35% and 45% respectively. This effect was explained by development of pronounced spike at zero in the eventwise ratio distributions.

The dynamical fluctuations of the p/π ratio follow scaling with number of accepted pions and protons, indicating that the signal originates from the resonance feed down. Concerning the scaling of the K/π ratio fluctuations, energy and centrality dependences of this observable behave differently as a function of number of kaons in the acceptance, indeed the centrality dependence is steeper than the energy dependence. Scaling of the K/p ratio fluctuations is the subject for futher studies.

Effect of the dE/dx resolution on the dynamical fluctuations of the K/π ratio was studied and it was found out that finite identification capabilities play role only in the very peripheral Pb + Pb collisions at $158A$ GeV beam energy and have no effect on the results for 20A GeV beam energy.

Concerning the origin of the K/π ratio fluctuations in the UrQMD model, the ϕ meson, decaying into pair of kaons, was proposed as the source of the positive fluctuations.

Chapter 6

The CBM experiment

The Compressed Baryonic Matter (CBM) is a future fixed target heavy-ion experiment at Facility for Antiproton and Ion Research (FAIR). Highest baryon densities will be created in A + A collisions at $10A$ - $35A$ GeV beam energy range. The goal of the experiment is to explore the properties of super dense nuclear matter, looking for in-medium modifications of hadrons, phase transition from dense hadronic matter to quark gluon plasma and for the critical point on the phase diagram of strongly interacting matter which is shown in figure 6.1.

At high temperatures and zero baryon chemical potential there is a region of crossover. The LHC expriments will investigate the phase diagram in this region. While the FAIR will explore the region of high baryon chemical potentials and moderate temperatures, where the first order phase transition occurs. This first order phase transition line ends with the critical point, existance and exact location of which is one of the subjects for the CBM experiment.

CBM will measure rare and penetrating probes such as dilepton pairs from light vector mesons and charmonium, open charm, multistrange hyperons together with collective hadron flow and fluctuations in heavy ion collisions.

Fundamental aspects of Quantum Chromodynamics and astrophysics will be covered in the program of the CBM experiment: the equation of state of strongly interacting matter at high baryon densties, the restoration of chiral symmetry, the origin of hadron masses, the confinement of quarks in hadrons, the structure of neutron stars, the dynamics of core-collapse of supernovae.

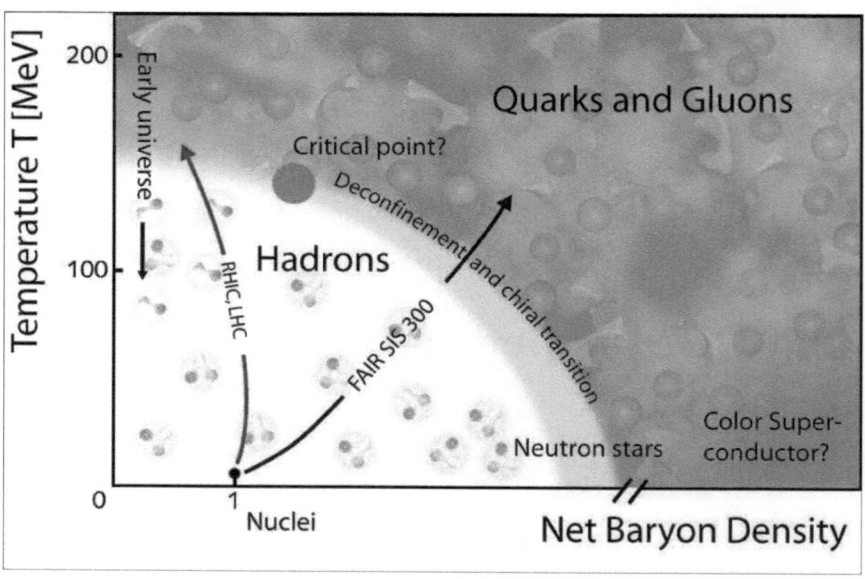

Figure 6.1: Sketch of the phase diagram of strongly interacting matter.

6.1 Detector design

The major experimental challenge for CBM is posed by the extremely high interaction rates of up to 10^7 events/second. These conditions require unprecedented detector performances concerning speed and radiation hardness. The detector signals are processed by a high-speed data acquisition and an online event selection system [47].

6.1.1 Micro vertex detector (MVD)

The Micro Vertex Detector will measure primary and secondary vertices with an ultimate precision. It consists of two detector planes based on Monolithic Active Pixel Sensor (MAPS) technology which are mounted in vacuum. With a small-size prototype a position resolution of 3μm was achieved. Current task of R&D is an improvement in read-out time and radiation hardness [48].

6.1. DETECTOR DESIGN

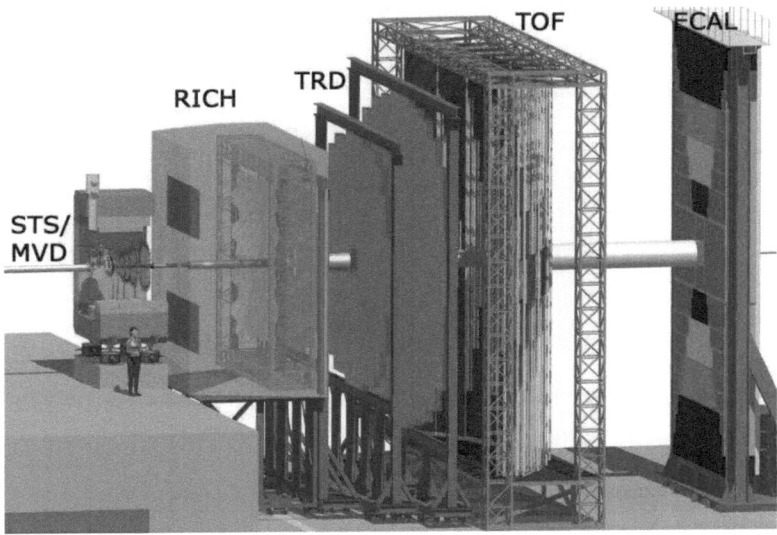

Figure 6.2: Sketch of the CBM detector. The beam comes from the left side. From left to right: Micro Vertex Detector and Silicon Tracker System inside of the dipole magnet, Ring Imaging Cherenkov detector, Transition Radiation Detectors, Time of Flight wall, Electromagnetic Calorimeter.

6.1.2 Silicon tracking system (STS)

The Silicon Tracking System (STS) serves for track measurement and for determination of primary and secondary vertices [49, 50, 51]. The current STS layout consists of minimum 8 layers (see figure 6.3) and is placed inside a magnetic dipole field which provides the bending power required for momentum determination with an accuracy of $\Delta p/p = 1\%$.

The STS has to fulfill the following requirements: material budget below 0.3% radiation length per layer to reduce multiple scattering, hit resolution of about 10 μm to achieve a vertex resolution of about 50 μm along the beam axis, radiation hardness up to a dose of 50 MRad corresponding to the dose accumulated in ten years of running and read-out times of less then 25 ns to accommodate reaction rates of 10 MHz.

One possible technology are Silicon microstrip detectors. The current layout foresees a pitch of 50 μm and three different strip lengths of 20, 40 and 60 mm. The strips are

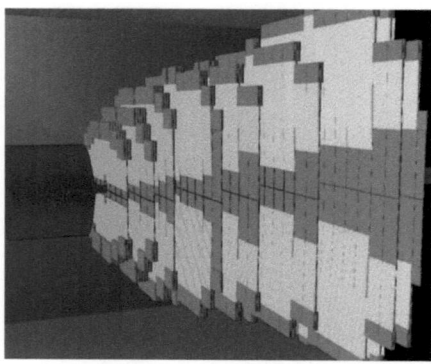

Figure 6.3: Sketch of the Silicon Tracking System. Tracking stations, support structure, cables and readout electronics are shown.

arranged such that the occupancy is below 2% for a central Au+Au collision at 25 AGeV. The detectors will be double sided with a stereo angle of 15° between the strips which has to be optimized by simulations

6.1.3 Ring imaging cherenkov (RICH) detectors

The RICH detector is designed to provide identification of electrons and suppression of pions in the momentum range of electrons from low-mass vector-meson decays. The actual layout of the RICH detector consists of a radiator, a mirror and a photon detector. The glass window of the photomultipliers is covered with wave-length shifter (WLS) films in order to increase the absorption of Cherenkov photons [52].

A cruicial task is to match the rings to the charged particle tracks. If the track position at the mirror can be determined with an accuracy of 200 μm, and assuming a momentum resolution of 1%, the mismatch of pions to electron rings is less than 10^{-3} per event. This number will be considerably improved when taking into account particle identification by time-of-flight measurement and by the TRD.

6.1.4 Transition radiation detector(TRD)

Three Transition Radiation Detector stations will serve for particle tracking and for the identification of high energy electrons and positrons ($\gamma > 2000$) which are used to recon-

6.1. DETECTOR DESIGN

struct J/ψ mesons. According to simulations which are based on the experience obtained with the development of the TRD for ALICE and of the TRT for ATLAS, pion suppression factors of up to 200 (for momenta above 2 GeV/c) at an electron efficiency of better than 90% can be achieved [53].

The major technical challenge is to develop highly granular and fast gaseous detectors which can stand the high rate environment of CBM in particular for the inner part of the detector planes covering forward emission angles. For example, at small forward angles and at a distance of 4 m from the target, we expect particle rates of more than 100 kHz/cm^2 for 10 MHz minimum bias Au+Au collisions at 25 AGeV. In a central collision, particle densities of about 0.05/cm^2 are reached. In order to keep the occupancy below 5% the size of the smallest types cell should be about 1 cm^2.

6.1.5 Resistive plate chambers (RPC)

An array of Resistive Plate Chambers will be used for hadron identification via TOF measurements. The TOF wall is located about 10 m downstream of the target and covers an active area of about 120 m^2. The required time resolution is about 80 ps. For 10 MHz minimum bias Au+Au collisions at 25 AGeV the innermost part of the detector has to work at rates up to 20 kHz/cm^2. At small deflection angles the pad size is about 5 cm^2 corresponding to an occupancy of below 5% for central Au+Au collisions at 25 AGeV. With a small-size prototype a time resolution of about 90 ps has been achieved at a rate of 25 kHz/cm^2. Future R&D concentrates on the rate capability, low resistivity material, long term stability and realization of large arrays with overall excellent timing performance [54].

6.1.6 Electromagnetic calorimeter (ECAL)

The elctromagnetic calorimeter will be used to measure direct photons, neutral mesons decaying into photons, electrons and muons. Simulations and R&D have been started based on shashlik type detector modules as used in HERA-B, PHENIX and LHCb [55]. Particular emphasis is put on a good energy resolution and a high pion suppression factor.

6.1.7 Muon chambers (MUCH)

An alternative option for the CBM experiment, namely muon measurement, is considered. It consists out of iron/carbon absorber layers with tracking stations in between, so called muon chambers [56]. This setup will not be discussed in details in the current work because there is no possibility for hadron measurement after the absorbers.

6.1.8 Projectile spectator detector (PSD)

The Projectile Spectator Detector (PSD) will measure the forward energy near the beam axis carried by projectile spectator nucleons and fragments. This measurement will allow to determine the number of nucleons participating in the nucleus-nucleus collision, and thus the collision centrality, on an event-by-event base. The detector concept is a compensating hadron calorimeter consisting of lead-scintillator sandwich modules with silicon photomultiplier light readout. A relative energy resolution of less than $50\%/\sqrt{E[GeV]}$ is aimed at.

6.2 Event reconstruction

The development of the current simulation and analysis framework has started at the end of 2003. The framework is completely ROOT based. The modified HADES geometry interface used in this framework enables the user to select (on the fly) between the new ROOT Geometry Modeler and the Geant3 native geometry to describe the detectors. The simulation is based on the Virtual Monte Carlo concept, which was developed by the ALICE collaboration and allows to select different engines (Geant3, Geant4, Fluka) for the transport of tracks. Moreover the analysis is organized using the ROOT Task mechanism.

The CBM experiment will collide heavy ions in the momentum range from $10A$ to $45A$ GeV ($Z/A = 0.5$) at 10 MHz interaction rate for rare probes. This means that tracking algorithms have to be not only efficient but also very fast in order to allow online event selection. In this section the track reconstruction in STS, TRD and global tracking will be discussed.

6.2. EVENT RECONSTRUCTION

6.2.1 STS tracking

STS is the second detector after the target (first station is placed at 30 cm after the target), thus track reconstruction in this device is a quite challenging task due to the high track and hit density and the nonhomogeneous magnetic field (see figure 6.4).

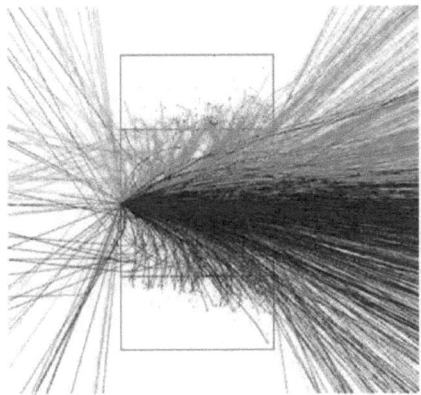

Figure 6.4: Visualization of the tracks, created in one central Au + Au collision at $25A$ GeV, inside the dipole magnet (1 m length).

In central Au + Au collision at $25A$ GeV about 600 - 700 tracks are accepted in the STS. A dedicated tracking algorithm, named Cellular Automaton (CA), was developed. The CA method creates short track segments in neighbouring detector planes and links them into tracks. Being essentially local and parallel the CA algorithm avoids exhaustive combinatorial searches. It internally uses a Kalman Filter for the track parameter propagation [57]. Note, that this algorithm requires four consequitive hits in order that a track can be reconstructed.

Here, the STS setup with double-sided micro strip detectors only was studied. Eight stations of the STS were considered. No charge sharing between the strips but simple Gassian smearing of the hit position in the STS detector was implemented. The cbmroot trunc version (revision number 6025) was used as a simulation tool.

Track finding efficiency as a function of momentum is shown in figure 6.5.

After finding a track, its parameters are determined using the Kalman Filter. The momentum resolution as a function of momentum is depicted in figure 6.6.

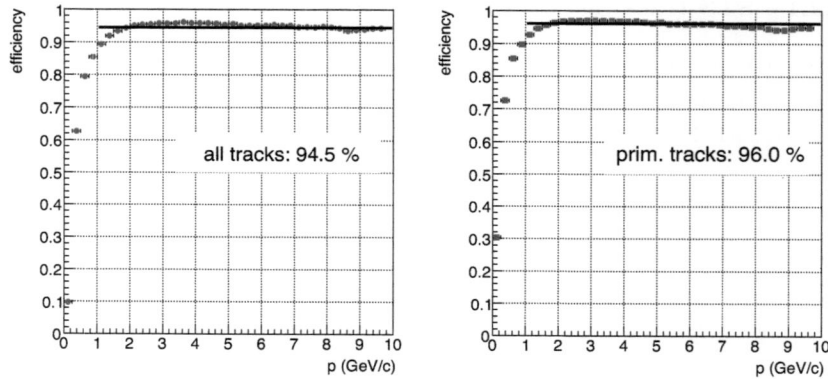

Figure 6.5: Efficiency of the track reconstruction in STS. Left picture - all tracks, right picture - primary vertex tracks (selected by cut on the track impact parameter). Solid line shows the average efficiency for the momentum range from 1 GeV/c to 10 GeV/c.

6.2.2 TRD tracking

The TRD setup with three stations each consisting out of four monolithic layers was used in this simulation. Realistic detector response was implemented by angular-dependent strip-like smearing of the hit positions with resolutions in the order of σ_x=300-500μm in the transverse and σ_y= 2.7-33mm in the longitudinal direction.

There are two approaches for the track reconstruction in the TRD: a standalone method which creates tracklets in different stations and connects them. Second, a 3D track following algorithm based on seeds from the STS. Tracks, reconstructed in the STS, are extrapolated to the first TRD station. Hits, which satisfy a searching criterium, are attached to the track. Afterwards track parameters are updated using the Kalman Filter and the track is propagated to the next station. Both methods show comparable efficiency, but the second is faster, because it does not need a combinatorial search. The efficiency of the track reconstruction as a function of momentum for the second approach is shown in figure 6.7. The efficiency drops down at lower momentum due to multiple scattering in the TRD material. The Advantage of the second method is that one does not need to merge STS and TRD tracks, this is directly included in the tracking itself. In the standalone method tracks need to be merged.

6.3. HADRON IDENTIFICATION

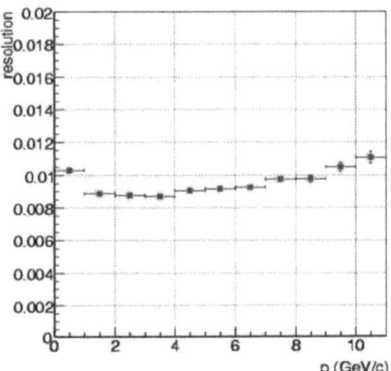

Figure 6.6: Momentum resolution as a function of momentum, obtained by fitting the tracks.

6.2.3 Global tracking

After the track is reconstructed in STS and TRD (for the standalone algorithm track segments also have to be merged) it is refitted using the Kalman Filter. Then track parameters at the last TRD station are propagated to the TOF wall and the closest TOF hit is attached to the track. Only one TOF hit can be attached to one global track. After merging with TOF is done, the track is refitted and the length of the trajectory is calculated starting from the primary vertex to the TOF hit.

The segmented pad like setup of the CBM TOF wall was implemented in the current simulation (with pad size 2x2 cm^2). The wall has eight gaps and produces a hit with realistic time responce out of eight Monte Carlo points created by a charged track during the simulation.

The efficiency of the global tracking is shown in figure 6.8. For tracks with $p > 1 GeV/c$ an efficiency of 86.4% is achieved.

6.3 Hadron identification

In the CBM experiment hadrons will be identified using a TOF wall, which is placed 10 m after the target. A simultaneous measurements of track the length l (assuming that the particle comes from the main vertex) and the time-of-flight t provides the velocity

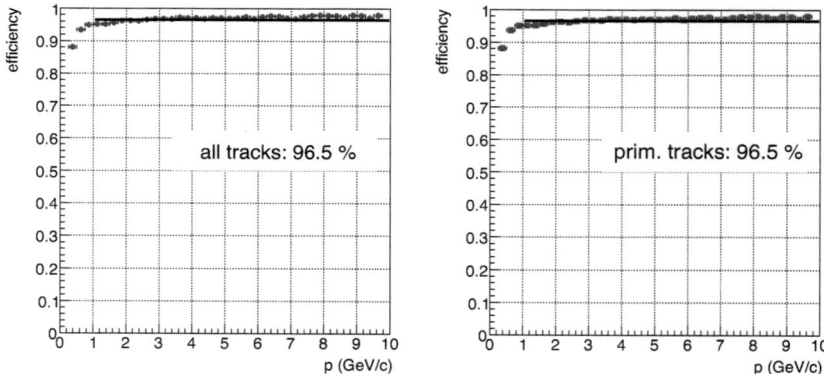

Figure 6.7: Efficiency of the track reconstruction in the TRD. Left picture - all tracks, right picture - primary vertex tracks. Solid line shows the average efficiency for momentum range from 1 GeV/c to 10 GeV/c.

of a particle $\beta = \frac{l}{ct}$. Knowing the momentum p from the track fit in the STS one can calculate the squared mass as

$$m^2 = p^2(\frac{1}{\beta^2} - 1) \tag{6.1}$$

If the time resolution σ_t dominates over uncertainties in momentum and track length measurements, the squared mass resolution is given by

$$\sigma_{m^2} = 2p^2 \frac{c^2 t}{l^2} \sigma_t \tag{6.2}$$

According to this, the squared mass resolution is proportional to the square of the momentum, thus at some momentum value the separation power will be limited and one will missidentify pions, kaons and protons.

Here we assume a time resolution of 80 ps. The squared mass spectrum of reconstructed particles in central Au + Au collisions at 25A GeV beam energy (simulated with UrQMD) is shown in figure 6.9.

For each momentum bin three Gaussians where fitted to the squared mass distribution. Since the squared mass resolution is independent on mass, the width of these peaks were assumed to be the same. The position of each peak was fixed to the value of the true

6.3. HADRON IDENTIFICATION

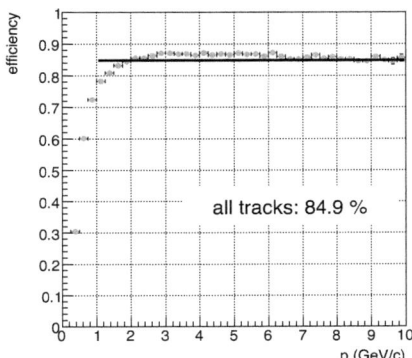

Figure 6.8: Global tracking efficiency. Includes tracking in STS, TRD and merging with TOF. The solid line shows the average efficiency for momentum range from 1 GeV/c to 10 GeV/c.

mass squared. So, in total there are 4 free parameters in the fit: the common width and three amplitudes. Only for low momenta (less then 2 GeV/c), where shift in mass due to energy loss becomes important, peak positions were released. The squared mass resolution (equation 6.2) as a function of momentum is shown in figure 6.10.

Since the aim is to study the eventwise kaon to pion ratio, kaons and pions have to be identified on a track-by-track basis. In order two identify kaons, a window around the true kaon mass squared was used. The initial width of the window is $\pm 2\sigma_{m^2}$. Due to the pion contamination the border next to pions of the window was shifted to the right to reach an overall purity of 50%. If the particle has a squared mass value that lies inside this window, it is identified as a kaon. The momentum distribution of all, accepted, reconstructed and identified K^+ is shown in figure 6.11.

Transverse momentum distribution at midrapidity of K^+ is shown in figure 6.12.

The phase space distribution of primary hadrons are shown in figure 6.13. Three cases are considered: all particles, accepted tracks, reconstructed and identified particles. Table 6.1 shows corresponding efficiencies for each particle type.

particle	geometrical acceptance (%)	reconstruction and identification efficiency (%)	total efficiency (%)
π^+	41.9	80.9	33.9
K^+	38.0	56.9	21.6
p	57.8	67.1	38.8

Table 6.1: Values of geometrical acceptance, reconstruction and identification efficiency for primary hadrons from central Au + Au collisions at $25A$ GeV beam energy.

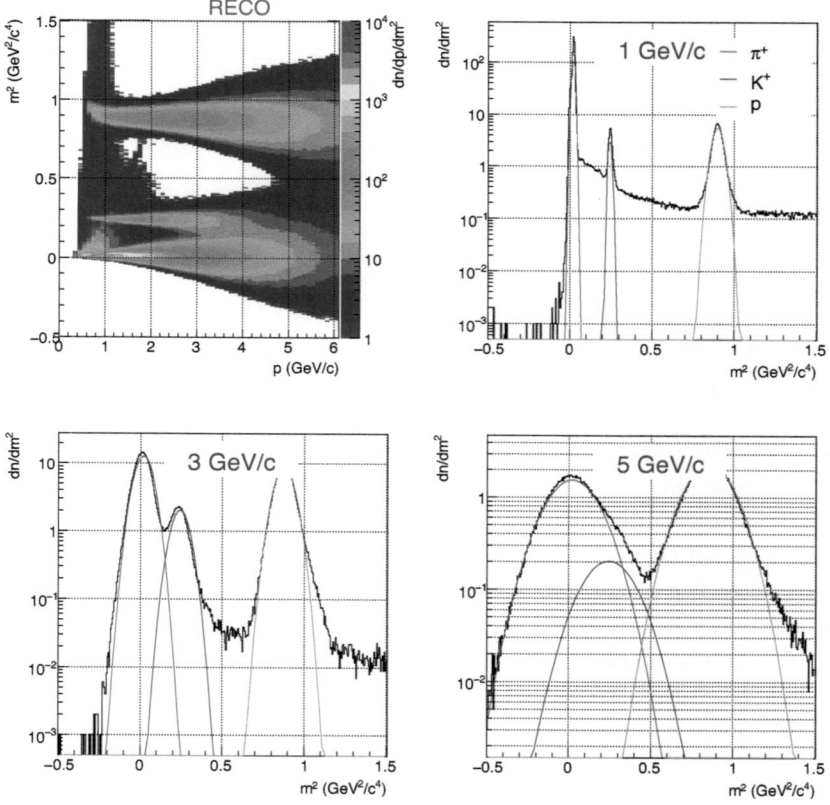

Figure 6.9: The squared mass spectrum of reconstructed particles. Projections for p = 1, 3 and 5 GeV/c are shown.

6.3. HADRON IDENTIFICATION

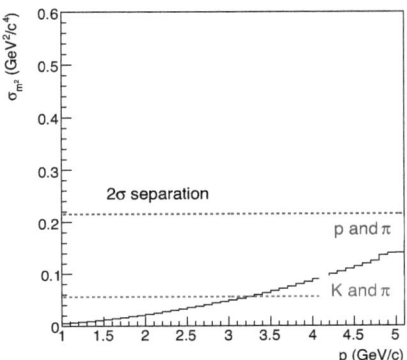

Figure 6.10: The squared mass resolution as a function of momentum. Two dashed horisontal lines show the border for a two sigma separation between kaons and pions (bottom), protons and pions (top).

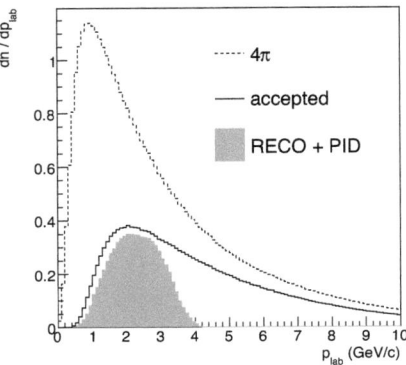

Figure 6.11: Momentum distribution of K^+ from central Au + Au collisions at $25A$ GeV beam energy: dashed line - all particles; solid line - accepted particles; filled region - reconstructed and identified tracks with purity of 50%.

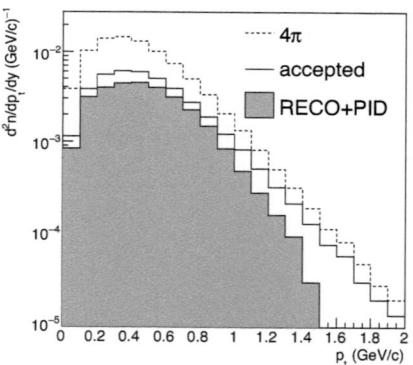

Figure 6.12: Transverse momentum distribution at midrapidity of K^+ from central Au + Au collisions at $25A$ GeV beam energy: dashed line - all particles; solid line - accepted particles; filled region - reconstructed and identified tracks with purity of 50%.

6.3. HADRON IDENTIFICATION 103

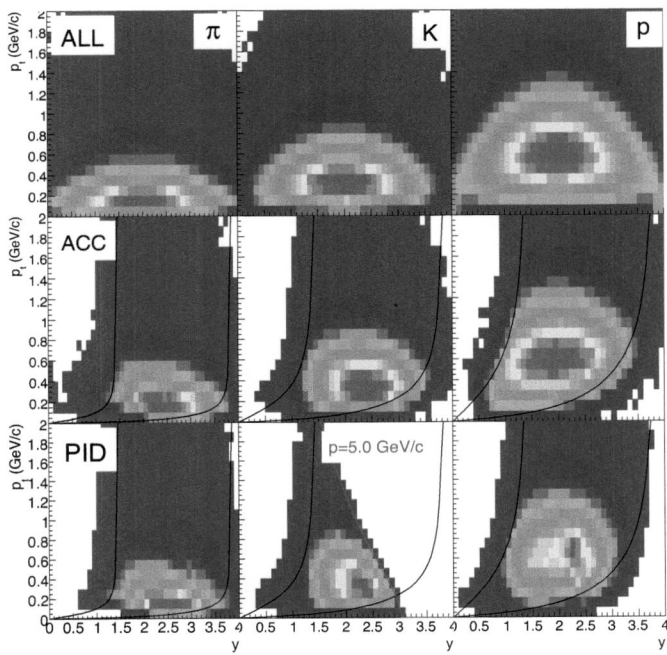

Figure 6.13: Phase space coverage of primary hadrons from central Au + Au collisions at $25A$ GeV beam energy: top row - all particles, middle row - accepted tracks, bottom row - reconstructed and identified particles (time resolution 80 ps, purity of kaon identification 50 %).

6.3.1 Hadron ID with CBM at SIS 100

In order to investigate the feasibility of hadron identification with CBM at SIS 100, central Au + Au collisions at 4A GeV beam energy were simulated with UrQMD and transported through the same setups of the STS and TOF as explained in the previous section. No scaling of the magnetic field was performed. No intermediate tracking was used in this study.

The squared mass distribution of reconstructed tracks is shown in figure 6.14.

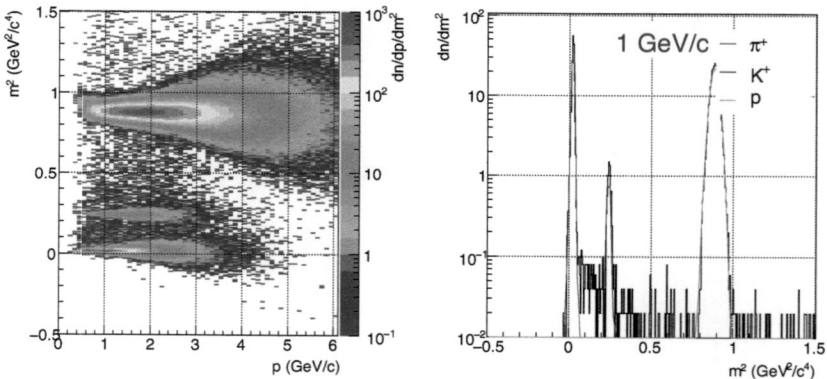

Figure 6.14: The squared mass spectrum of reconstructed tracks from central Au + Au collisions at 4A GeV beam energy. Projection for p = 1 GeV/c is shown.

The phase space coverage of primary hadrons from central Au + Au collisions at 4A GeV beam energy is shown in figure 6.15.

Transverse momentum distribution at midrapidity of K^+ is shown in figure 6.16.

In the light version of the CBM setup, one might consider moving TOF wall closer to the interaction point (approximately 4 meters), in order to detect kaons before they decay.

6.3. HADRON IDENTIFICATION

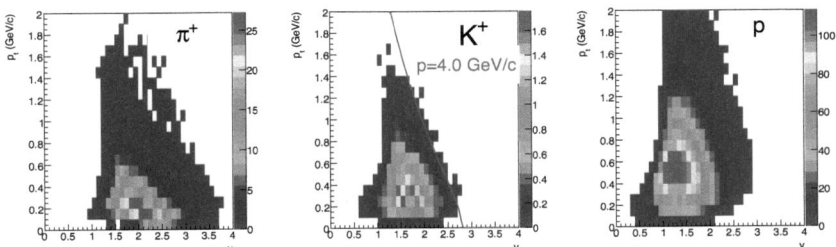

Figure 6.15: Phase space coverage of reconstructed and identified primary tracks from central Au + Au collisions at $4A$ GeV beam energy.

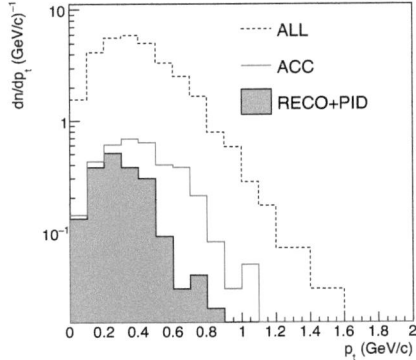

Figure 6.16: Transverse momentum distribution at midrapidity of K^+ from central Au + Au collisions at $4A$ GeV beam energy.

6.4 Event-by-event fluctuations of the kaon to pion yield ratio

The dynamical fluctuations of the kaon to pion yield ratio were extracted as was explained in the introduction chapter.

6.4.1 Simulations with UrQMD

Simulations, described here, were done with UrQMD v1.3 for central (b = 0 fm) Au + Au collisions at $25A$ GeV.

The dependence of the relative width of same and mixed events kaon to pion ratio distributions on the average number of kaons in the acceptance is shown in figure 6.17. The increase of σ_{same} towards lower number of accepted kaons is explained by the finite number statistics and is reproduced in the mixed events, so this effect can be corrected for.

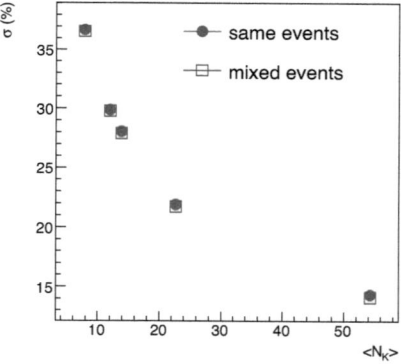

Figure 6.17: Relative width of eventwise kaon to pion ratio distribution for same events as a function of average number of kaons in the acceptance.

The purity of the kaon identification plays an important role in the analysis of dynamical fluctuations since it introduces limitations on the momentum range of identified particles. We have studied the effect of purity on dynamical fluctuations of the kaon to pion ratio (see figure 6.19). In addition we applied momentum cut (which corresponds to certain purity) in case of MC identification (see figure 6.18 for corresponding momentum cut values).

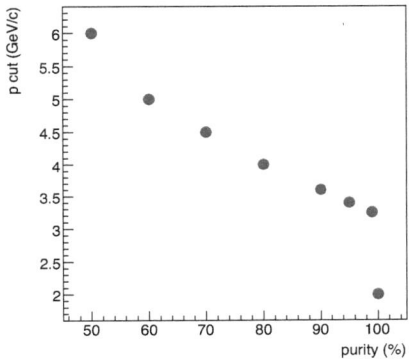

Figure 6.18: Dependence of the upper momentum cut on the required purity of the kaon identification.

Both real and MC identification show comparable results and indicate an increase of dynamical fluctuations when limiting the momentum interval. The important conclusion is that particle identification procedure does not introduce any significant bias to the dynamical fluctuations of the kaon to pion ratio.

Distributions of the event-by-event kaon to pion yield ratio for all particles (4π) and reconstructed and identified tracks is shown in figure 6.20. CBM setup as explained in context of the hadron identification.

Extracted dynamical fluctuations in 4π: $(2.6\pm0.2)\%$, after reconstruction and identification with purity of 50%: $(2.8\pm0.7)\%$. The interesting fact is that dynamical fluctuations are present in UrQMD and they are positive. Influence of ghost tracks and misidentification on the dynamical fluctuations was studied. Results in case of considering only correctly reconstructed and identified tracks is shown in figure 6.21. The dynamical fluctuations are $(2.8\pm0.6)\%$.

The values in 4π acceptance and after reconstruction and identification are comparable within the statistical errors. Ghost tracks introduce an increase of the dynamical fluctuations, thus we have systematic error of 0.2%.

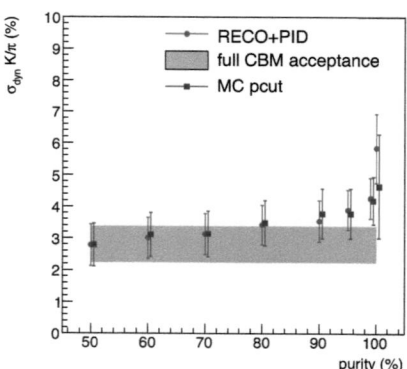

Figure 6.19: Dynamical fluctuations of the kaon to pion ratio as a function of purity of kaon identification.

For comparison, eventwise distributions of the kaon to pion yield ratio for same and mixed events for the purity of kaon identification 99% is shown in figure 6.22.

In contrast to the NA49 experiment, there is no spike at zero developed in the distribution even for such an extreme case as 99% purity of identification, which corresponds to the upper momentum cut of 3.2 GeV/c.

6.4. EVENT-BY-EVENT FLUCTUATIONS OF THE KAON TO PION YIELD RATIO

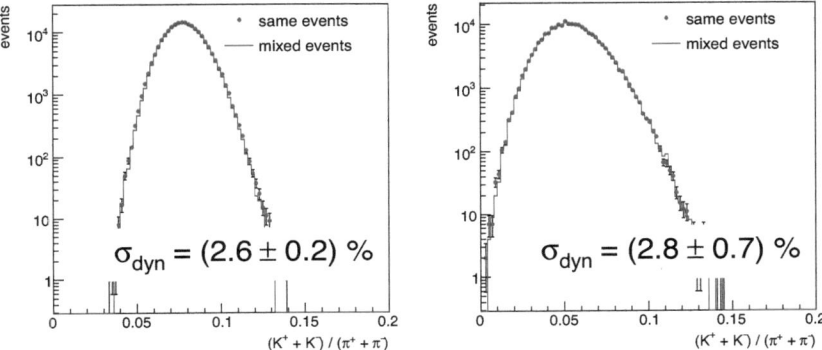

Figure 6.20: Distribution of the event-by-event kaon to pion yield ratio for same (points) and mixed (histogram) events. Left picture - all particles, right picture - reconstructed and identified (with 50% purity) tracks.

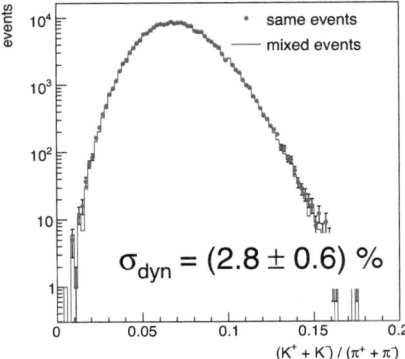

Figure 6.21: Distribution of the event-by-event kaon to pion yield ratio for same (points) and mixed (histogram) events for correctly reconstructed tracks using MC identification.

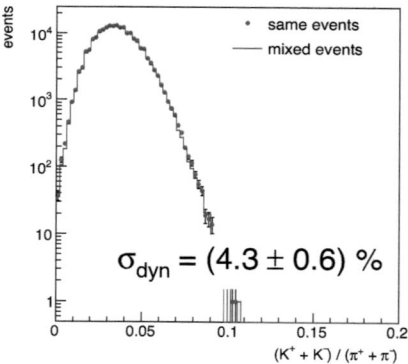

Figure 6.22: Distribution of the event-by-event kaon to pion yield ratio for same (points) and mixed (histogram) events for purity of kaon identification 99%.

6.4.2 Sensitivity test with Toy Model

Since the origin of positive dynamical fluctuations of the kaon to pion yield ratio in UrQMD is unknown we have created a simple Toy Model which simulates fluctuations of the kaon number for a sensitivity test. This Toy Model is described in chapter 5. Here, yields and kinematics of pions, kaons and protons are taken from UrQMD. Producing the bulk of independently created particles. Additional fluctuations of the kaon number of a different amplitude are added (see figure 6.23).

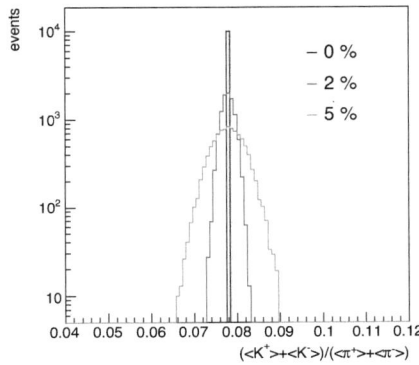

Figure 6.23: Fluctuations of the kaon number per event of different amplitudes.

These events are processed through the simulation and reconstruction chain of CBM. The extracted values of dynamical fluctuations after reconstruction and particle identification are compared to values in 4π. This comparison is shown in figure 6.24.

The conclusion is that by the reconstruction and identification we do not introduce a bias to the dynamical fluctuations, and CBM is sensitive to fluctuations on the level of 1%. This may be caused by the fact that kaon number fluctuations have uniform distribution in momentum.

6.5 Conclusion

The feasibility of measurement of dynamical fluctuations of particle yield ratios with proposed CBM setup was shown. Such effects as acceptance, reconstruction efficiency and influence of purity of the kaon identification were studied in details. No bias to

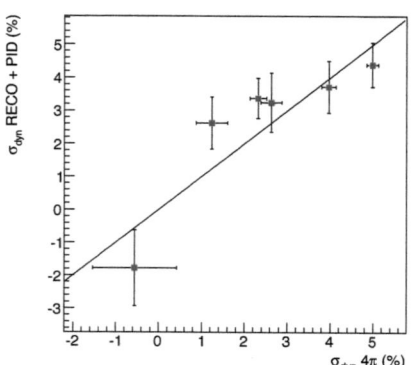

Figure 6.24: Sensitivity test with the Toy Model. Dynamical fluctuations extracted after reconstruction and identification versus dynamical fluctuations for Monte Carlo identification. Solid line shows ideal response.

dynamical fluctuations due to identification procedure was observed.

Sensitivity test with Toy Model shows that CBM is sensitive to fluctuations on the level of 1%.

CBM will be able to provide a high quality data on particle ratio fluctuations in heavy ion collisions since in contrast to NA49 it has symmetric azimuthal angle acceptance and will be designed to perform measurements at the energy range from $10A$ to $40A$ GeV with approximately the same acceptance for different particle species (see table 6.2).

Beam energy (AGeV)	π (%)	K (%)	p (%)
15	33.74	29.01	51.47
25	37.04	31.20	47.96
35	38.17	31.19	44.65

Table 6.2: The values of geometrical acceptance of different particle types for different beam energies.

Appendix A

Dependence on centrality bin size

A.1 Central events

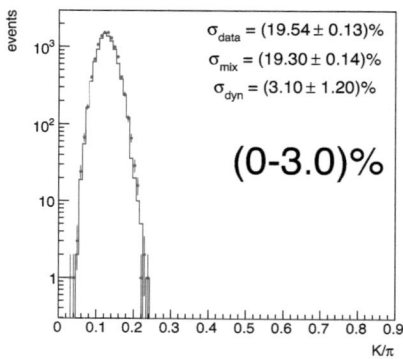
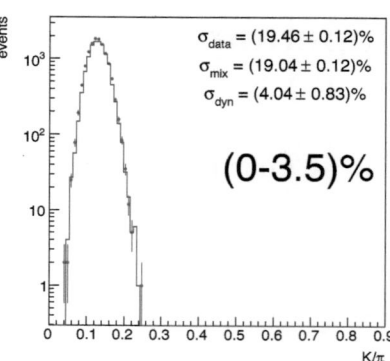
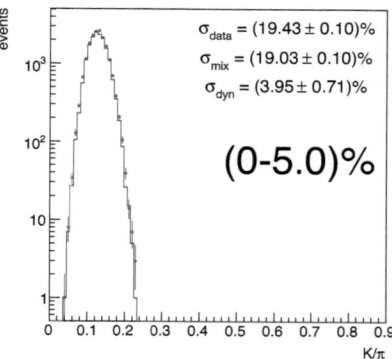
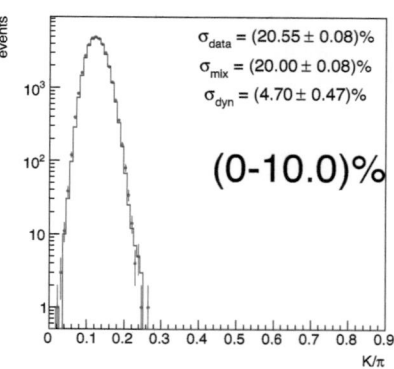

A.1. CENTRAL EVENTS

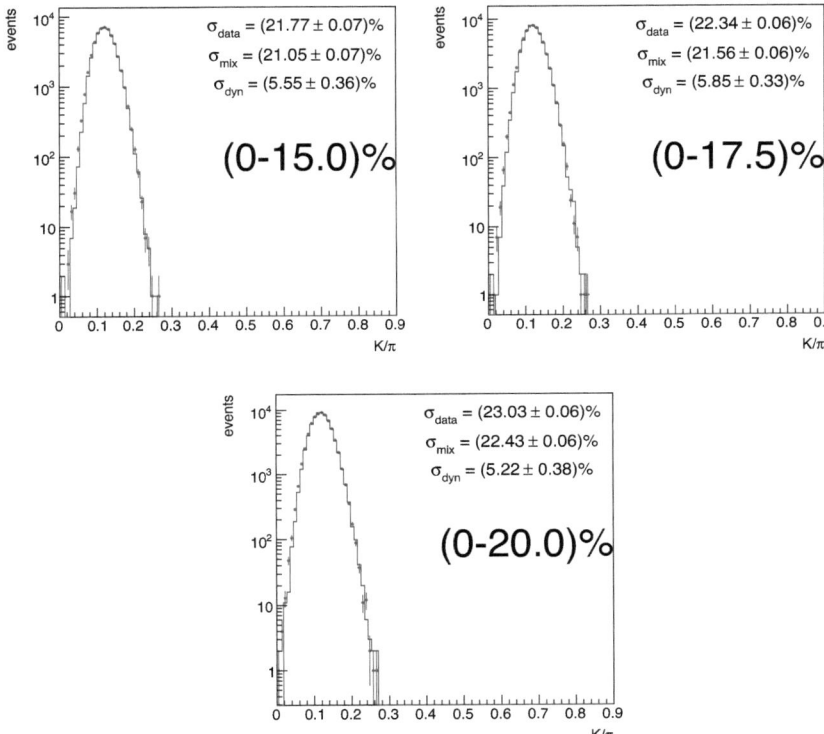

Figure A.1: Distributions of the event-by-event kaon to pion yield ratio for data (points) and mixed (histogram) events for different centrality bin widths as indicated in the figure for central Pb + Pb collisions at $158A$ GeV (data set 01J).

APPENDIX A. DEPENDENCE ON CENTRALITY BIN SIZE

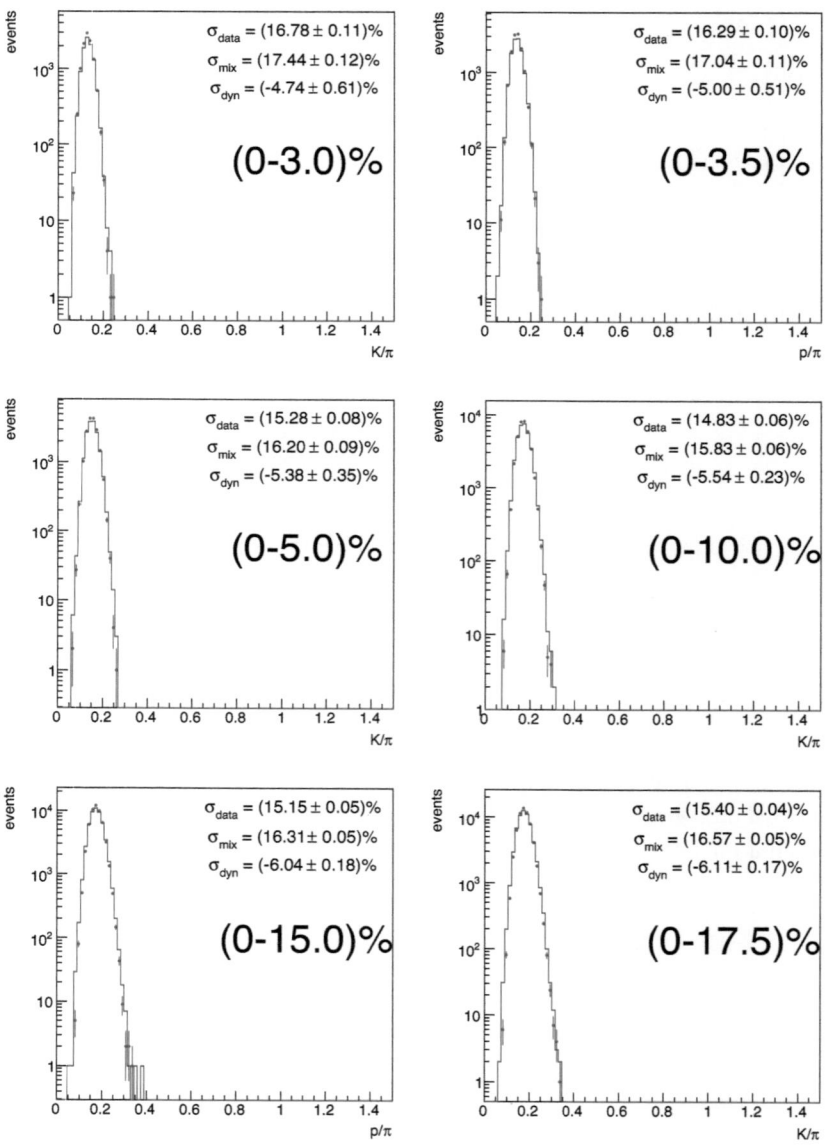

A.2. SEMI-PERIPHERAL EVENTS

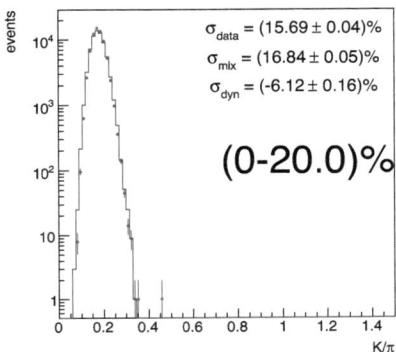

Figure A.2: Distributions of the event-by-event proton to pion yield ratio for data (points) and mixed events (histogram) for different centrality bin width for central Pb + Pb collisions at $158A$ GeV (data set 01J).

A.2 Semi-peripheral events

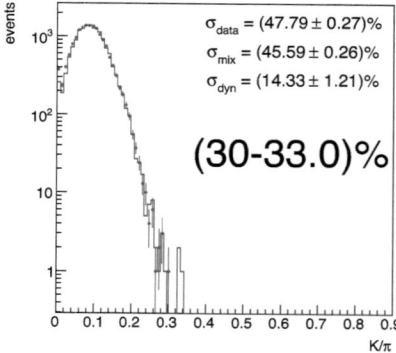
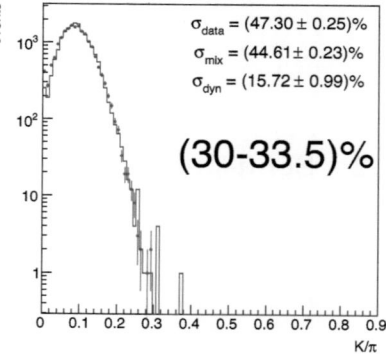

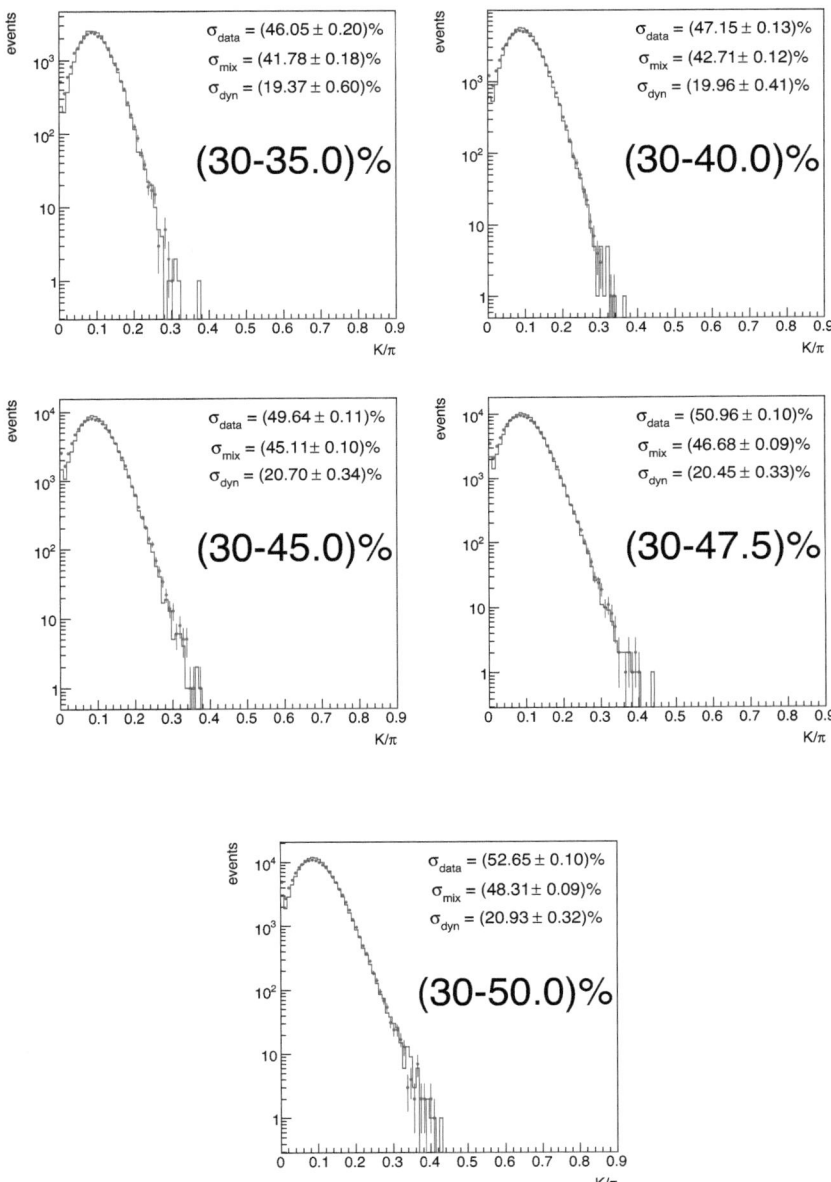

Figure A.3: Distributions of the event-by-event kaon to pion ratio for data (points) and mixed events (histogram) for different centrality bin width for semi-peripheral Pb + Pb collisions at 158A GeV beam energy (data set 01J).

APPENDIX A. DEPENDENCE ON CENTRALITY BIN SIZE

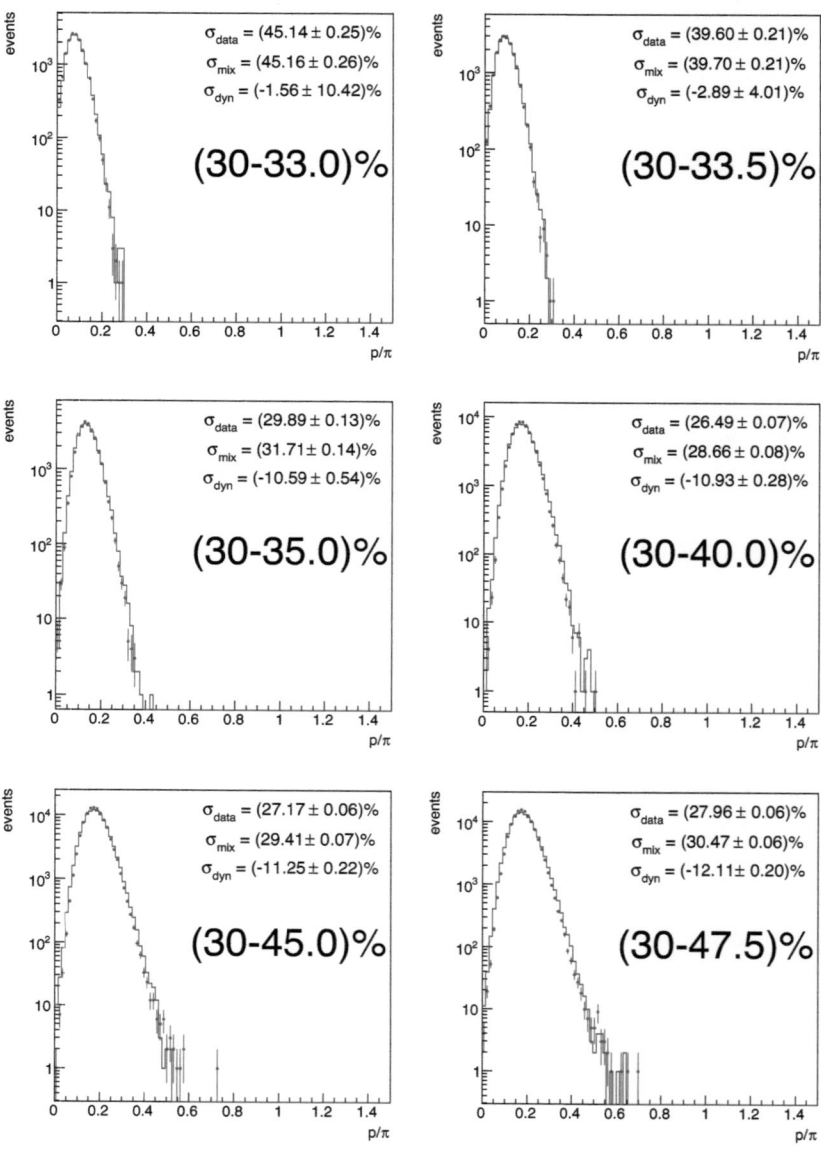

A.2. SEMI-PERIPHERAL EVENTS

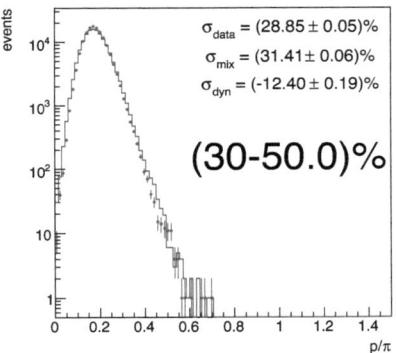

Figure A.4: Distributions of the event-by-event proton to pion ratio for the data (points) and mixed (histogram) events for different centrality bin width for semi-peripheral Pb + Pb collisions.

Appendix B

Centrality dependence

B.1 5% bin width

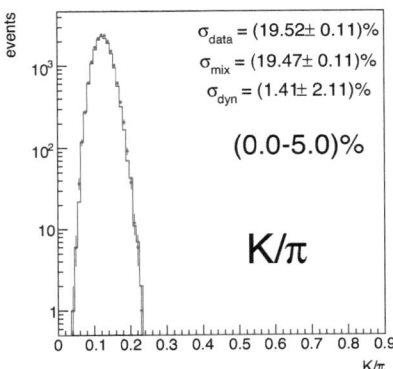
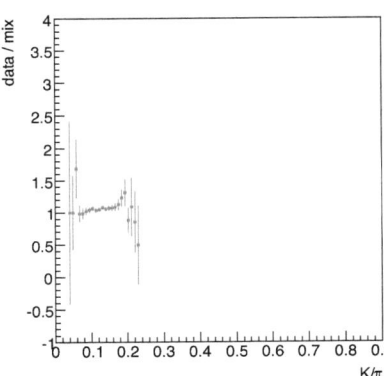

APPENDIX B. CENTRALITY DEPENDENCE

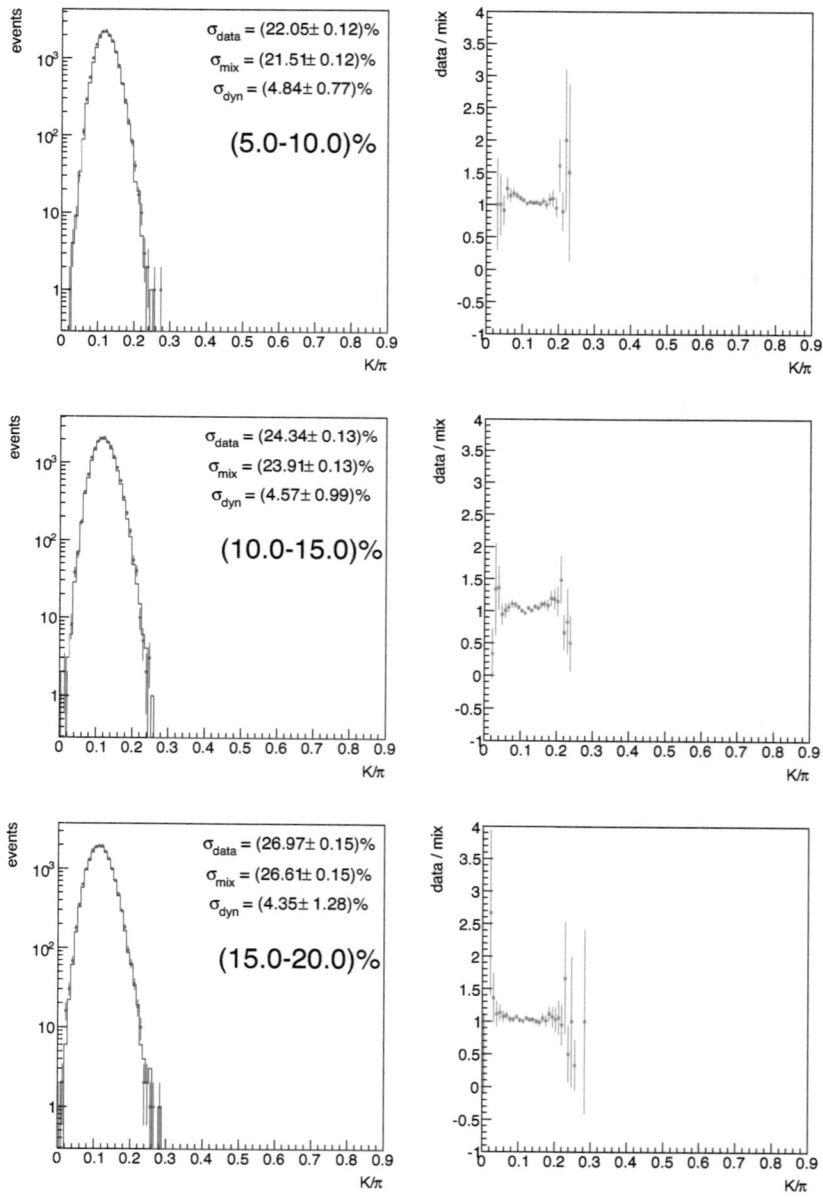

B.1. 5% BIN WIDTH

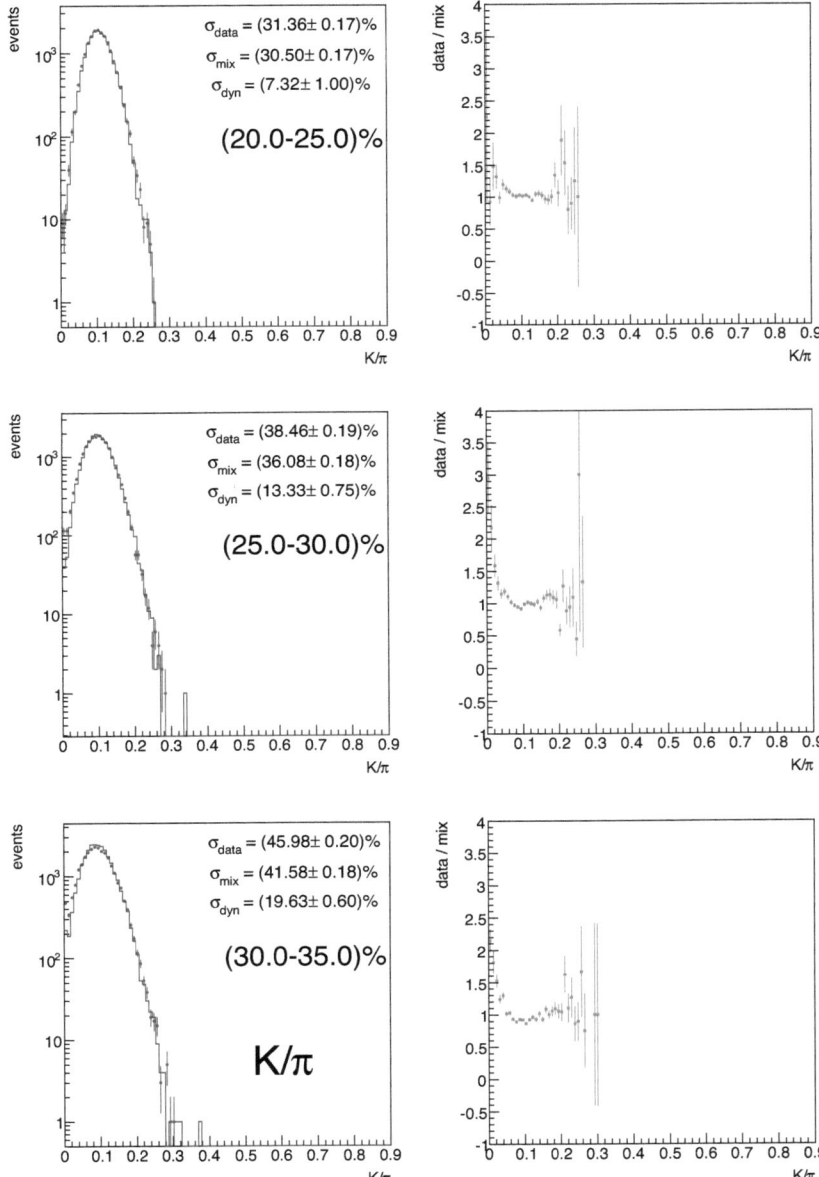

APPENDIX B. CENTRALITY DEPENDENCE

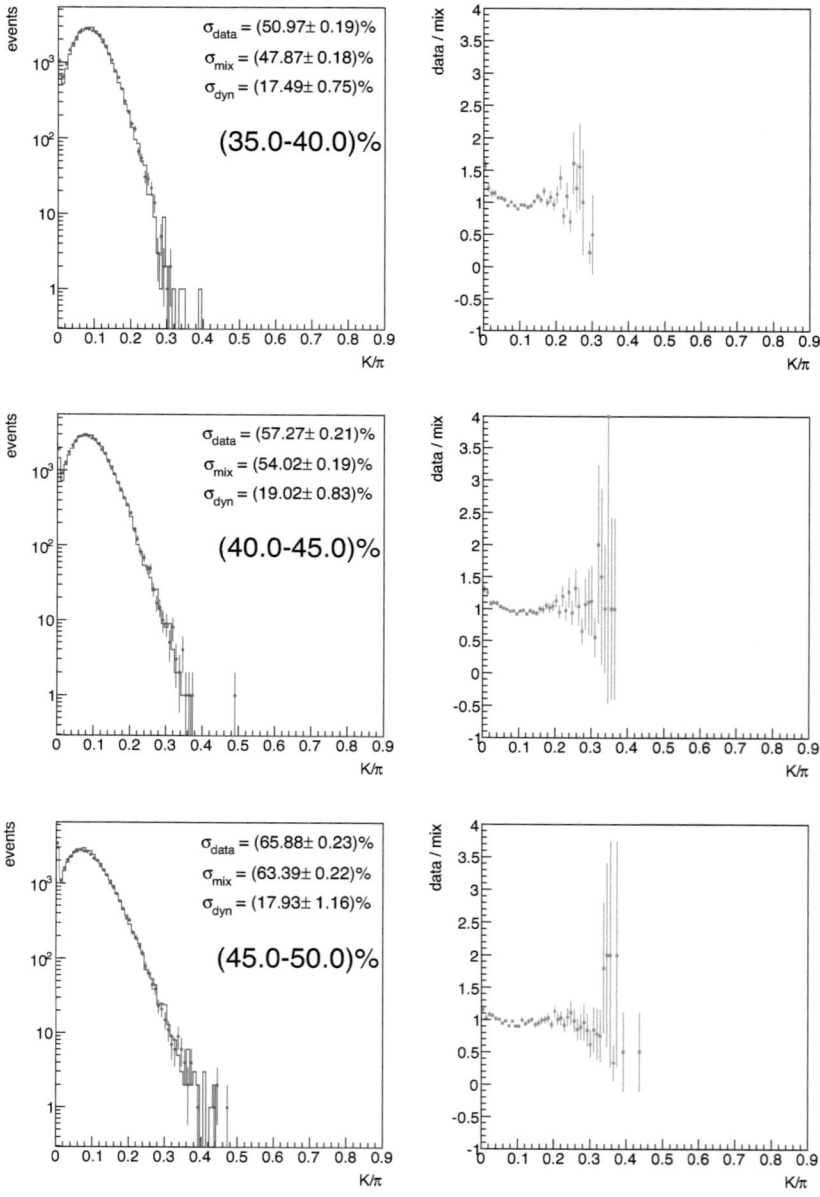

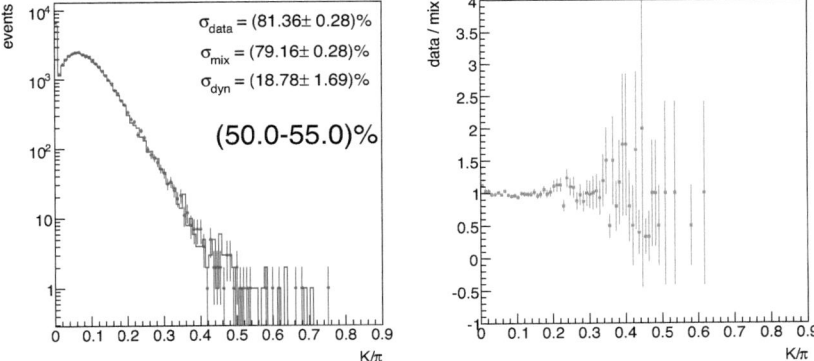

Figure B.1: Distributions of the event-by-event K/π ratio for data and mixed events for different centrality bins of minimum bias Pb + Pb collisions at $158A$ GeV. Right panels show the ratio of data to mixed events.

APPENDIX B. CENTRALITY DEPENDENCE

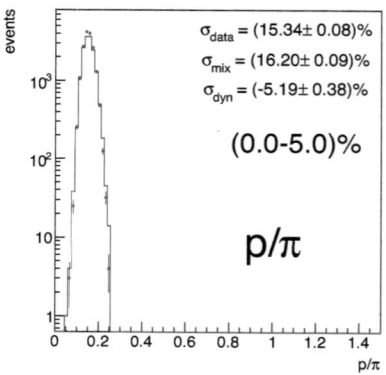
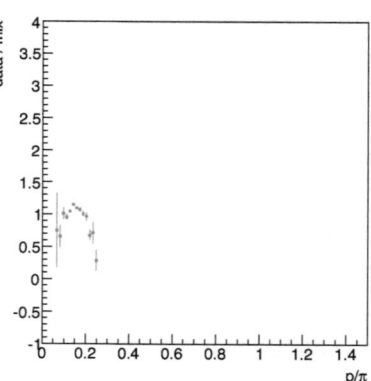
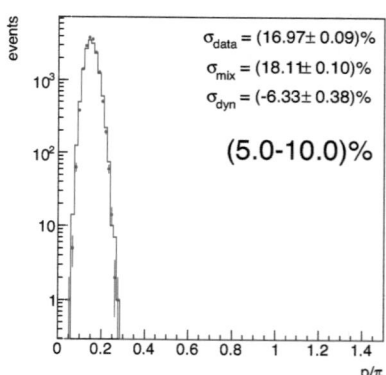
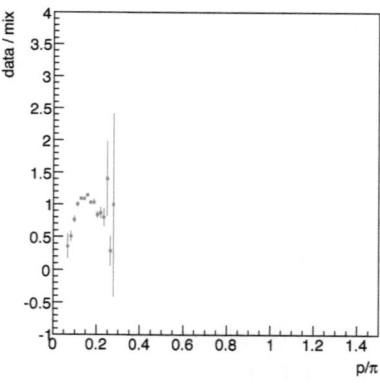

B.1. 5% BIN WIDTH

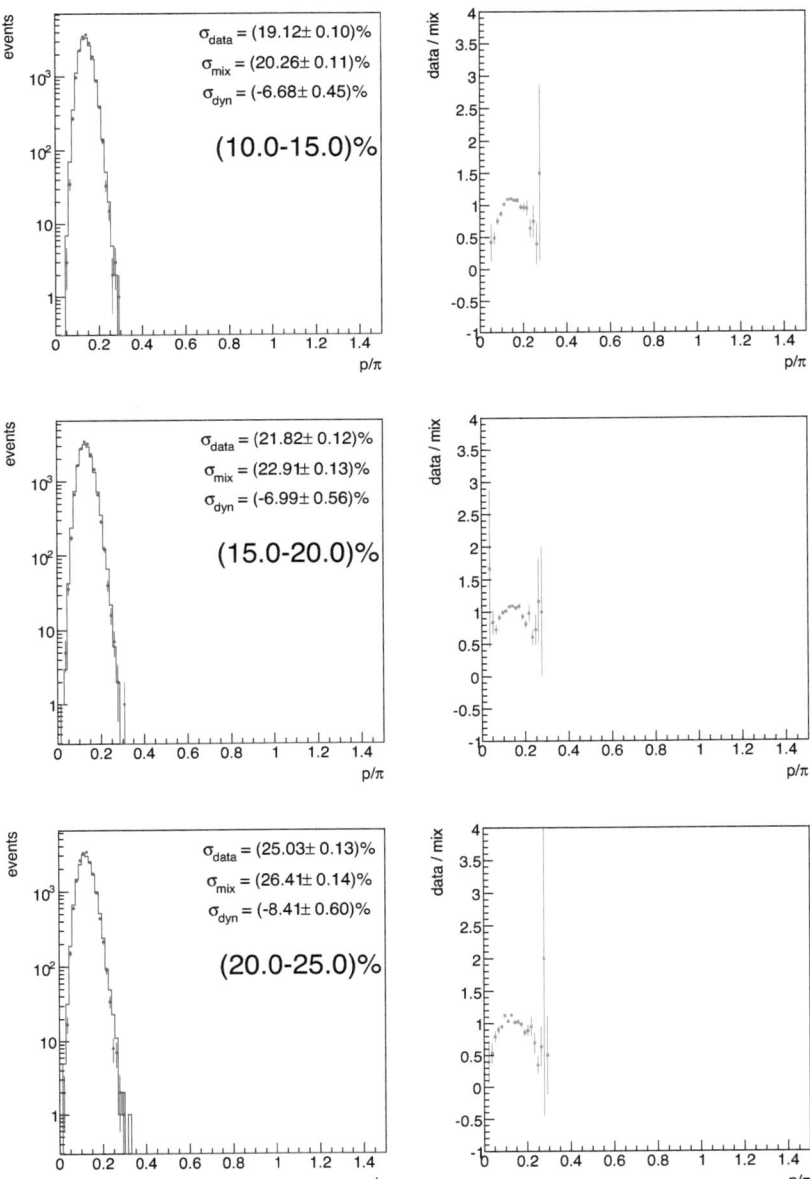

APPENDIX B. CENTRALITY DEPENDENCE

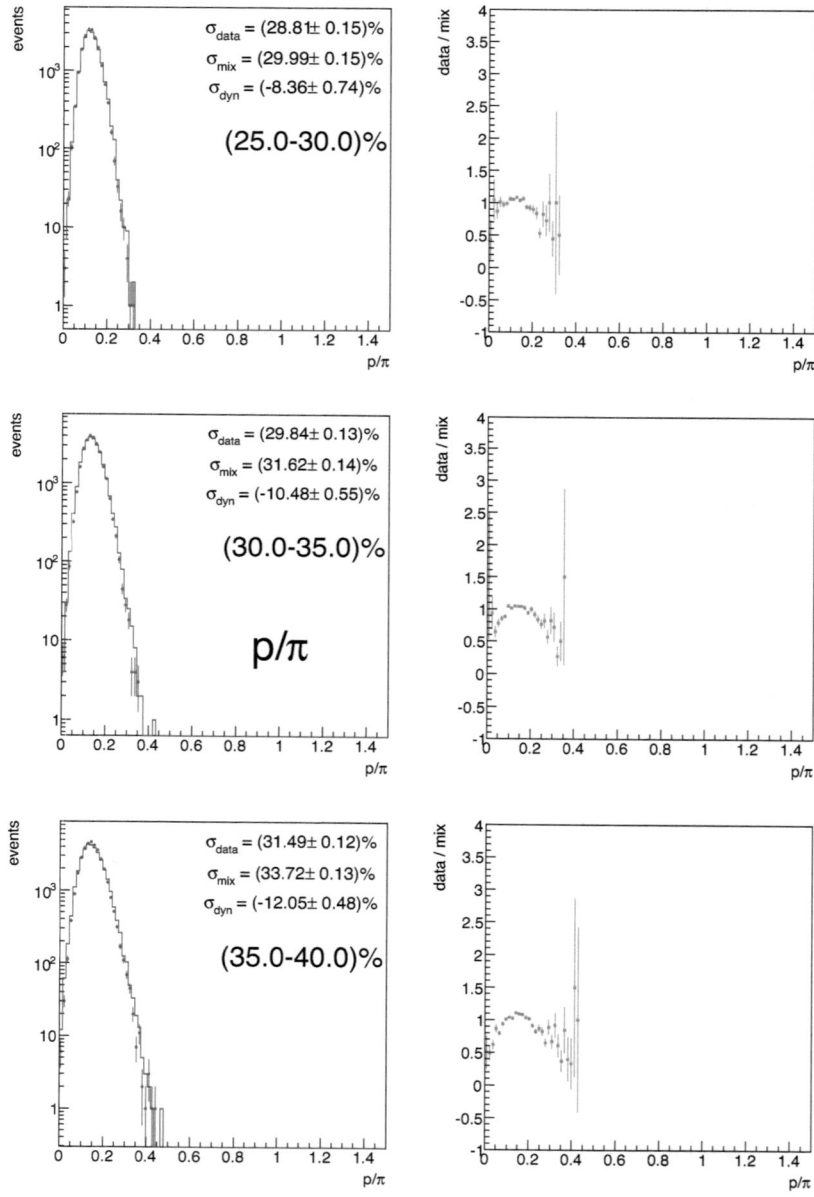

B.1. 5% BIN WIDTH

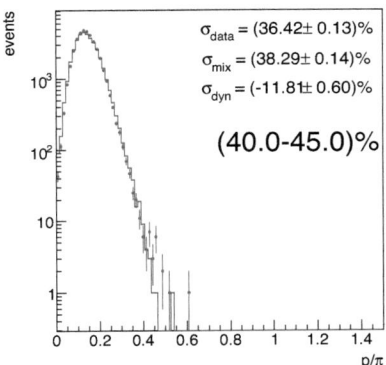
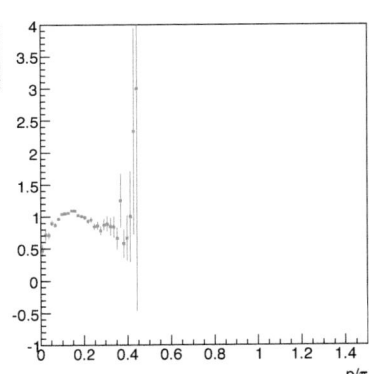

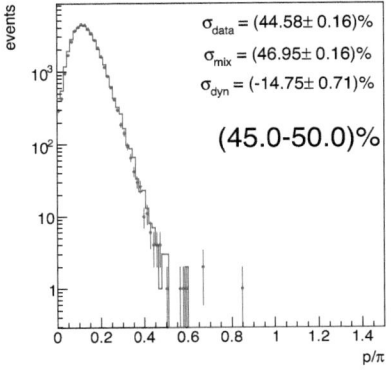
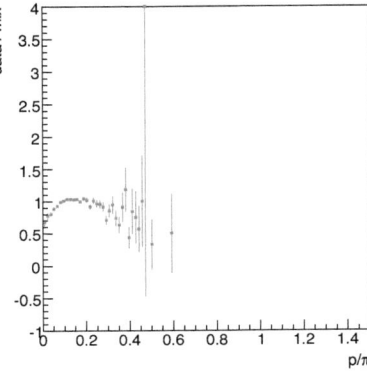

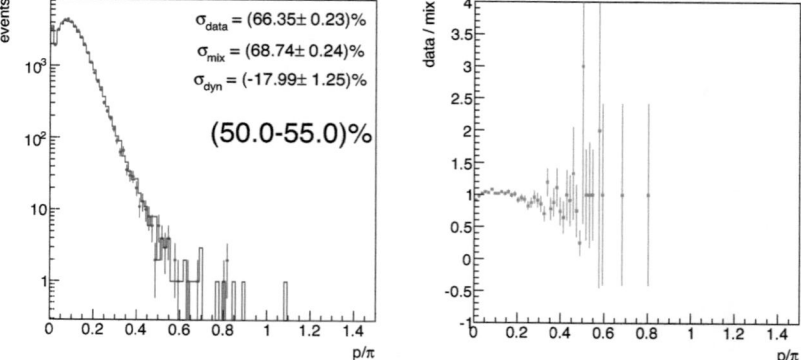

Figure B.2: Distributions of the event-by-event p/π ratio for data and mixed events for different centrality bins of minimum bias Pb + Pb collisions at 158A GeV. Right panels show the ratio of data to mixed events.

B.1. 5% BIN WIDTH

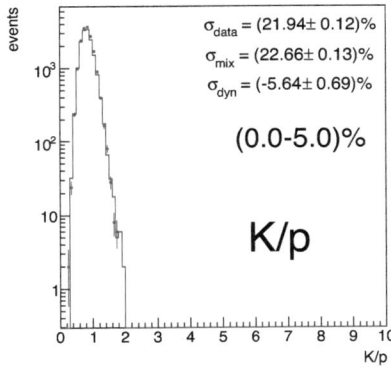
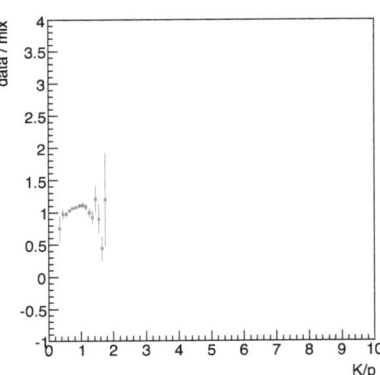
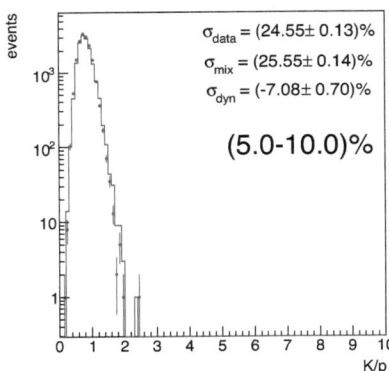
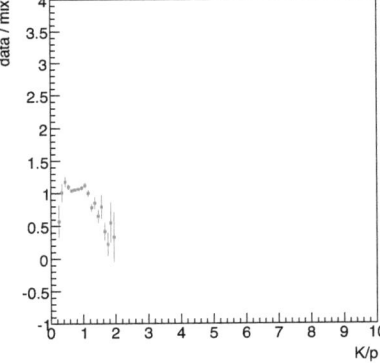

APPENDIX B. CENTRALITY DEPENDENCE

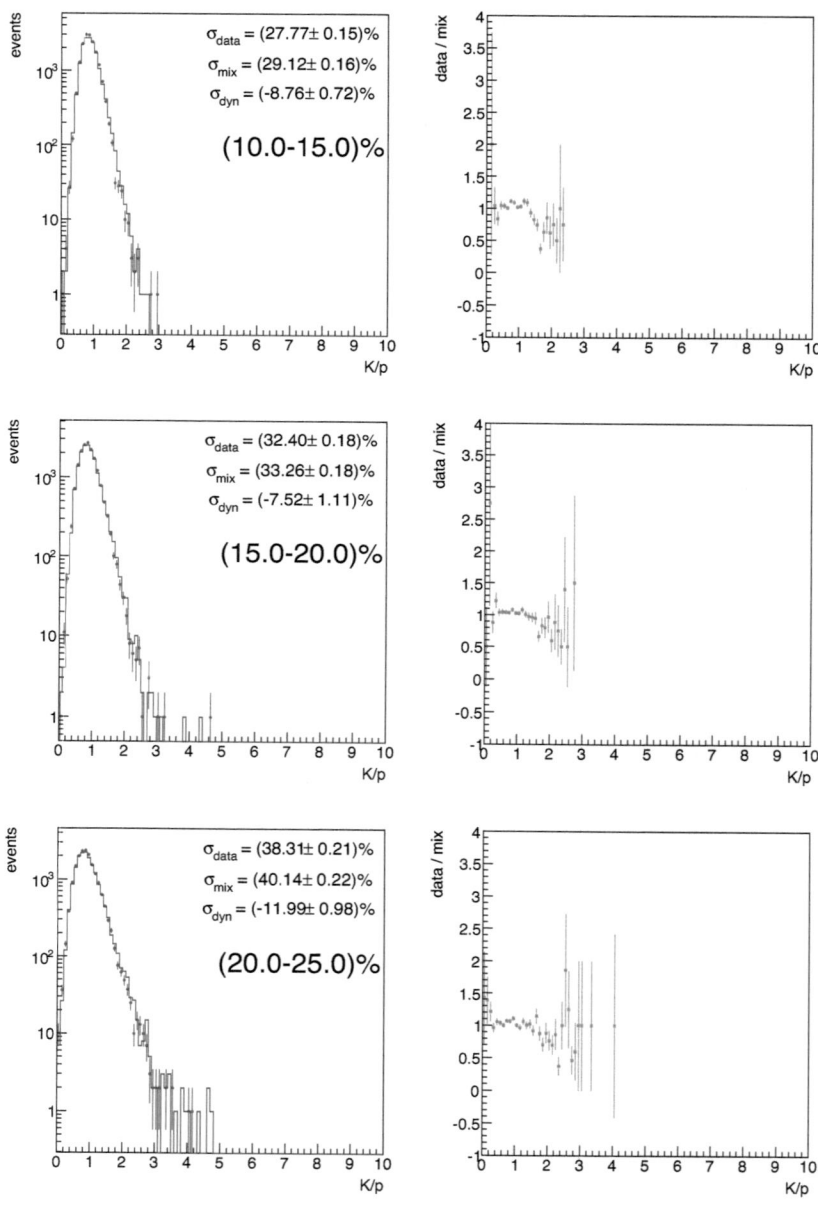

B.1. 5% BIN WIDTH

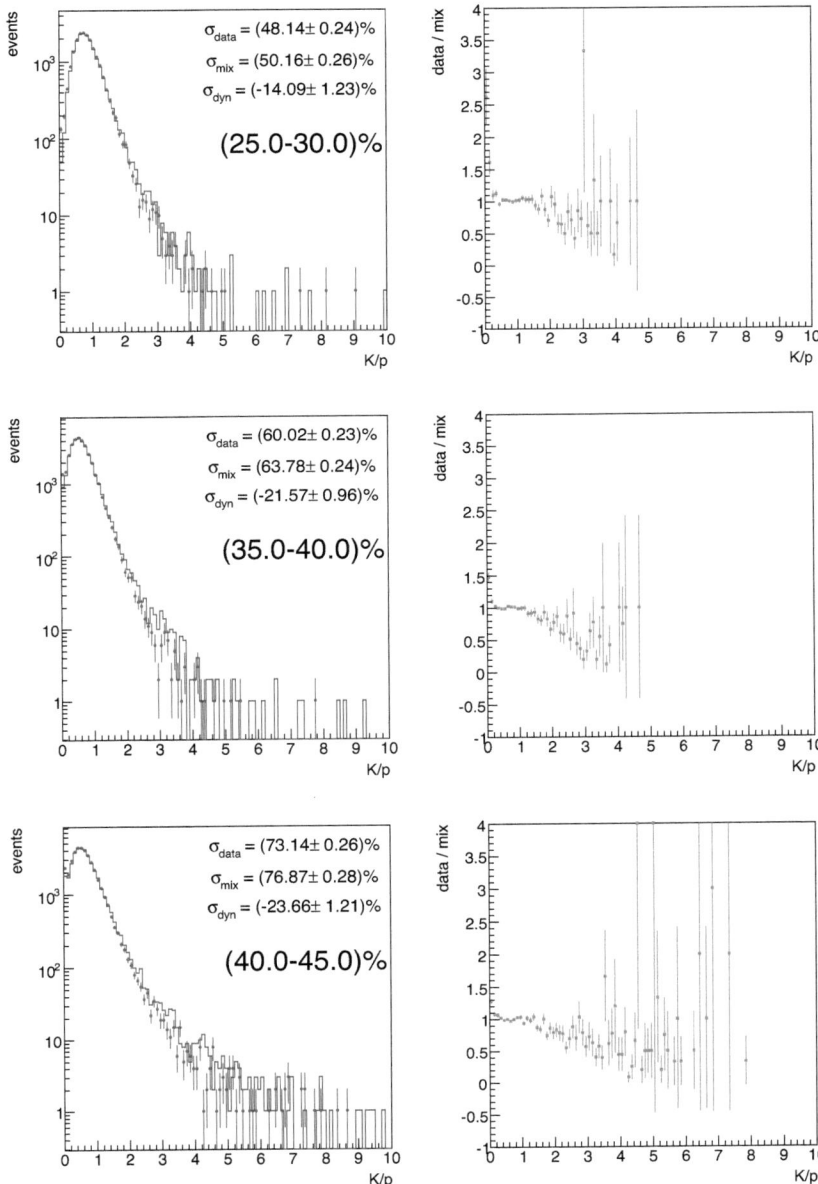

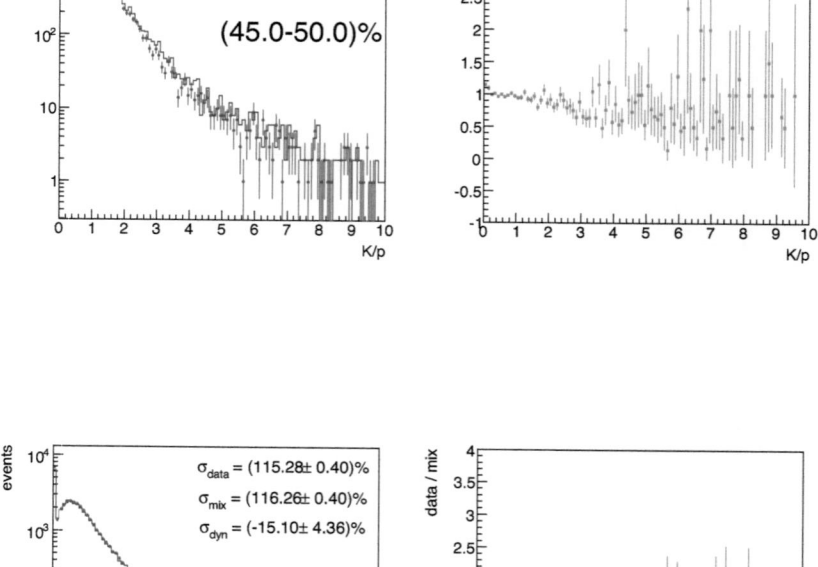

Figure B.3: Distributions of the event-by-event K/p ratio for data and mixed events for different centrality bins of minimum bias Pb + Pb collisions at 158A GeV. Right panels show the ratio of data to mixed events.

B.2 10% bin width

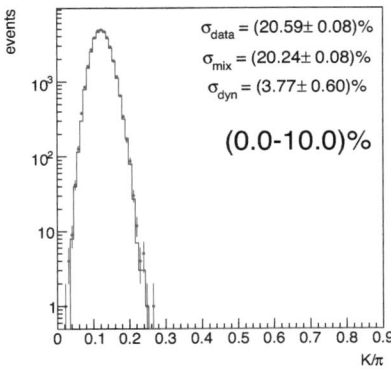
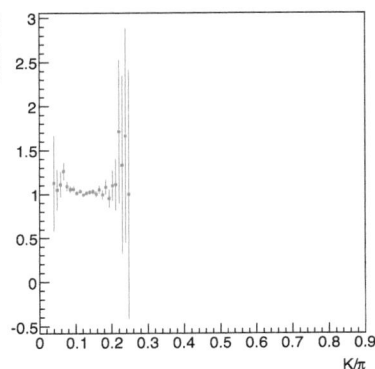
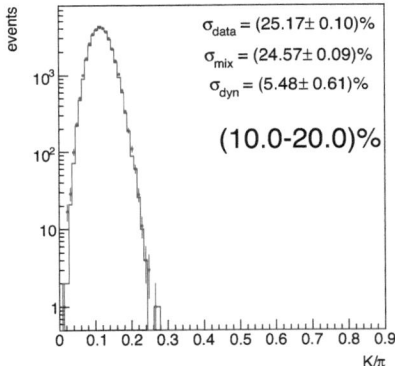
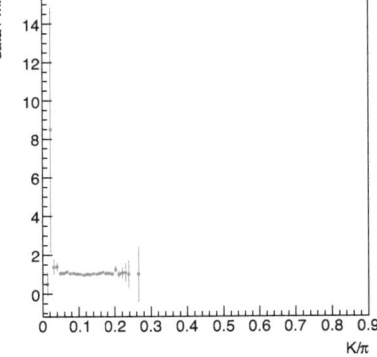

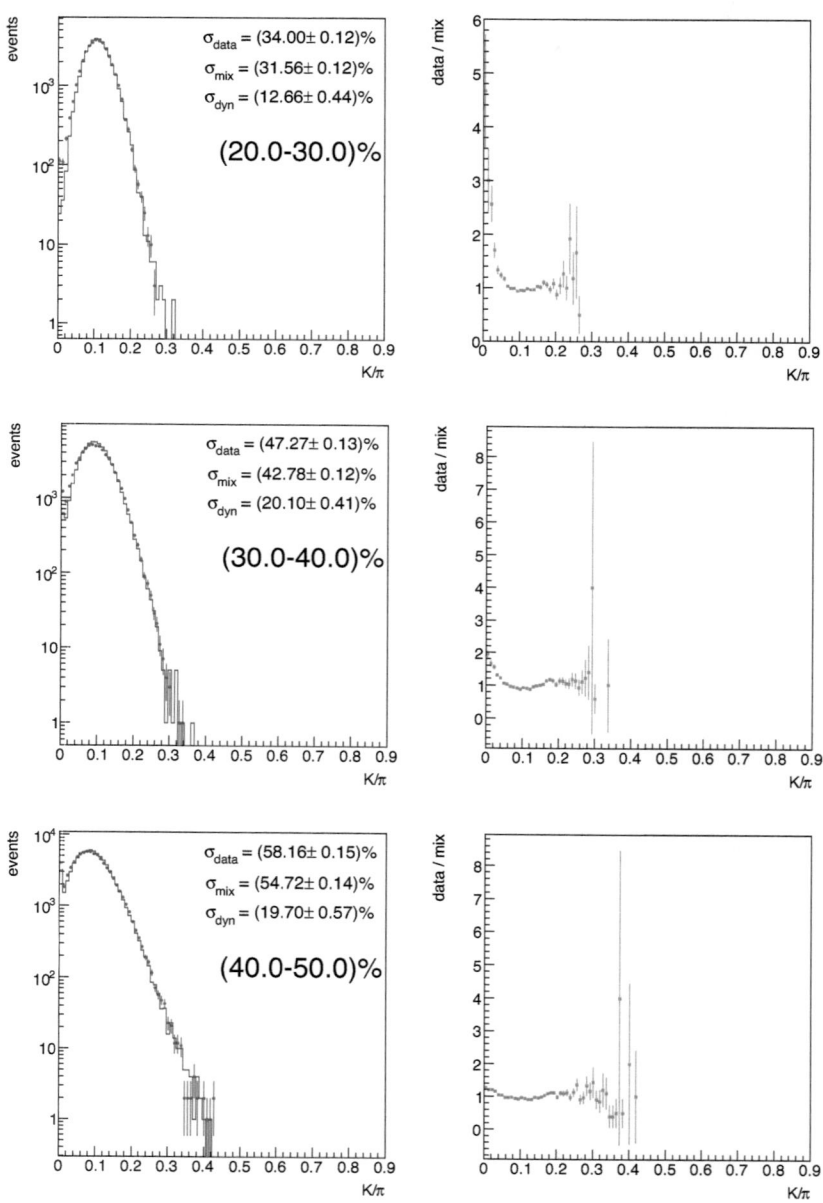

B.2. 10% BIN WIDTH

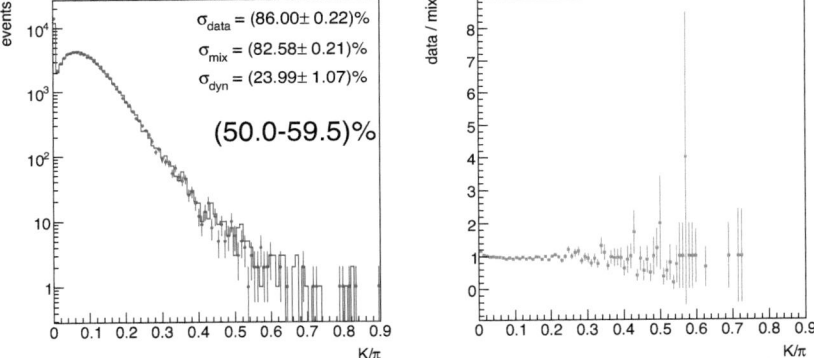

Figure B.4: Distributions of the event-by-event K/π ratio for data and mixed events for 10% centrality bins of minimum bias Pb + Pb collisions at 158A GeV. Right panels show the ratio of data to mixed events.

APPENDIX B. CENTRALITY DEPENDENCE

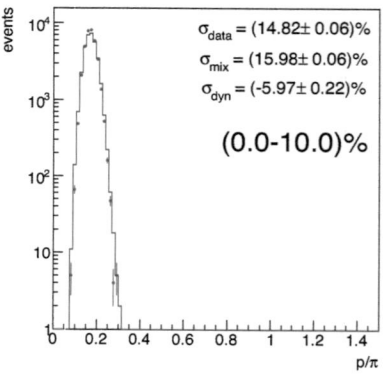
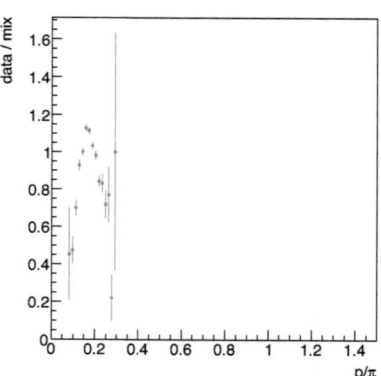
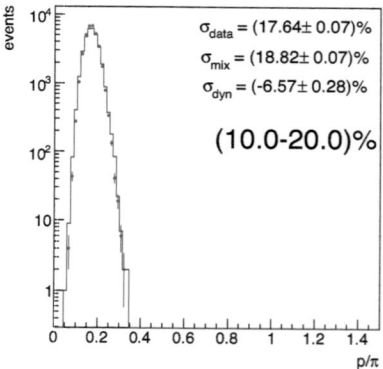
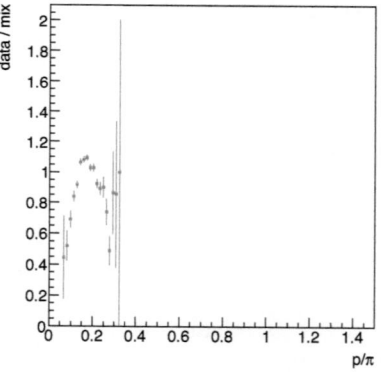

B.2. 10% BIN WIDTH

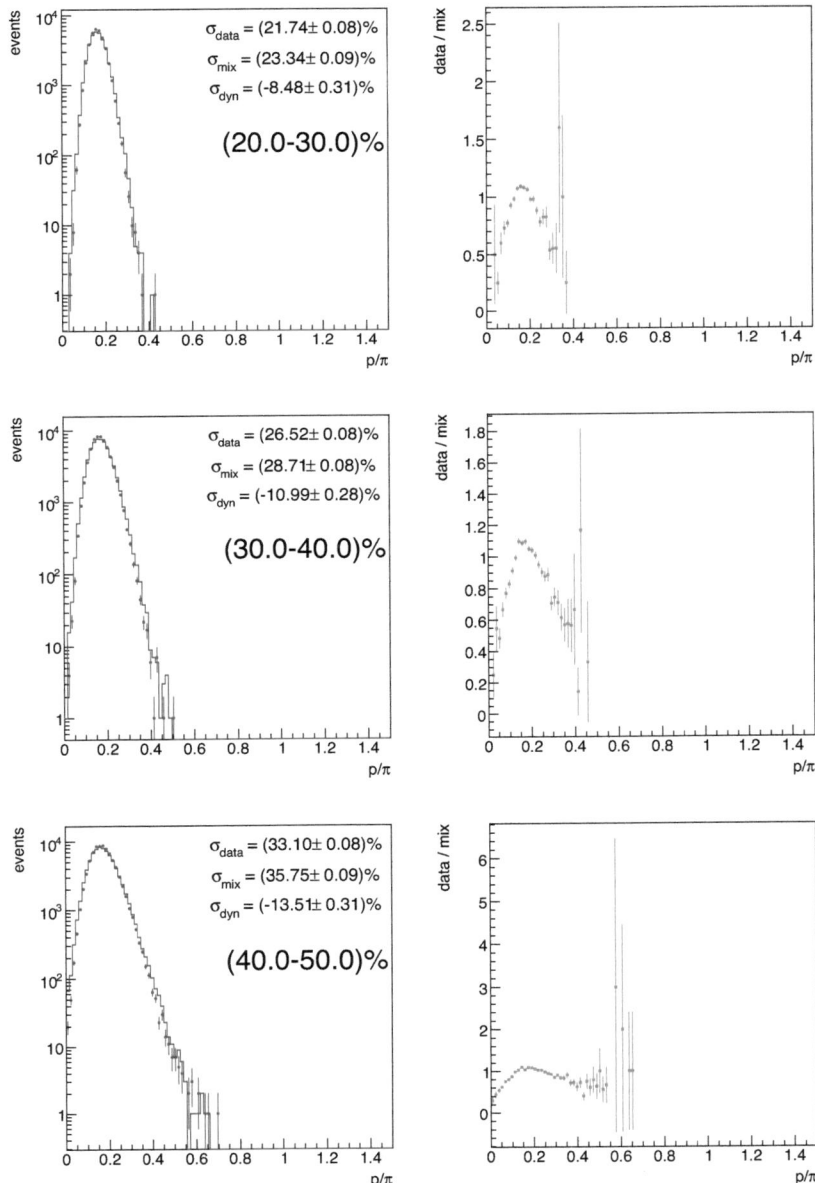

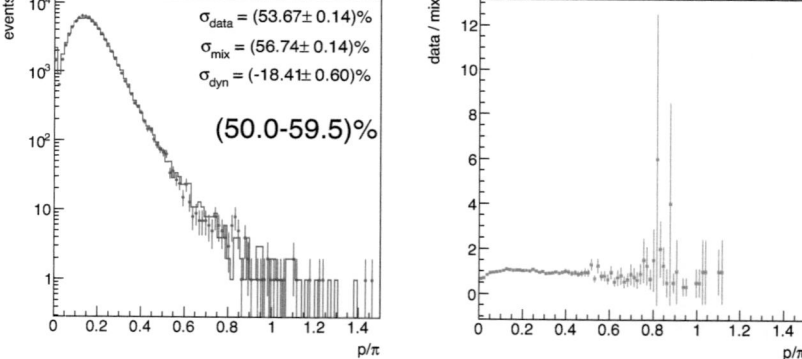

Figure B.5: Distributions of the event-by-event p/π ratio for data and mixed events for 10% centrality bins of minimum bias Pb + Pb collisions at 158A GeV. Right panels show the ratio of data to mixed events.

B.2. 10% BIN WIDTH

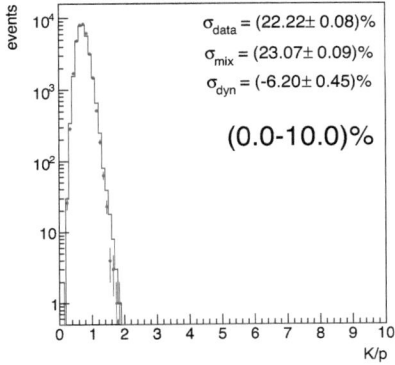
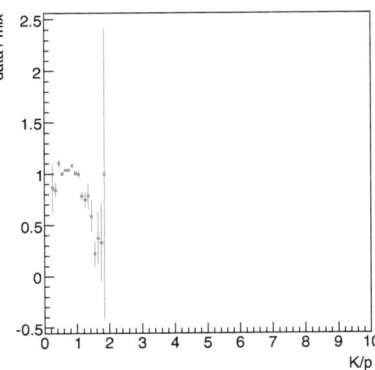
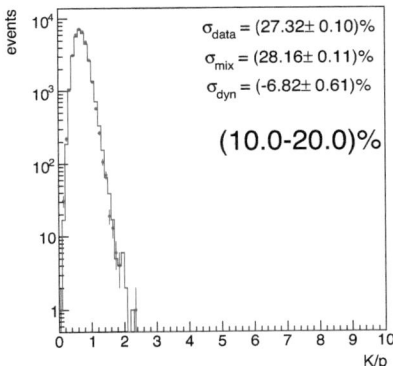
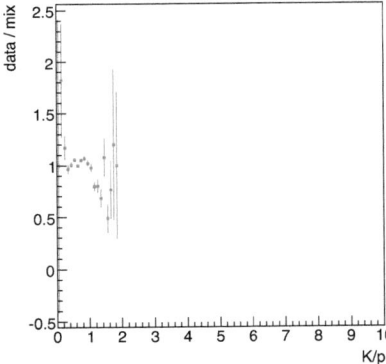

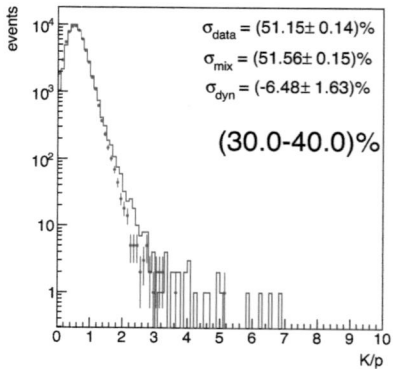
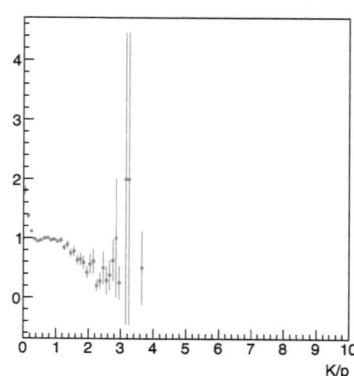

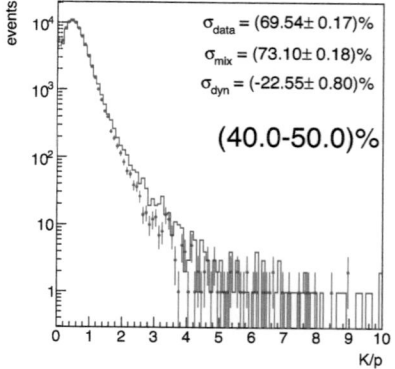
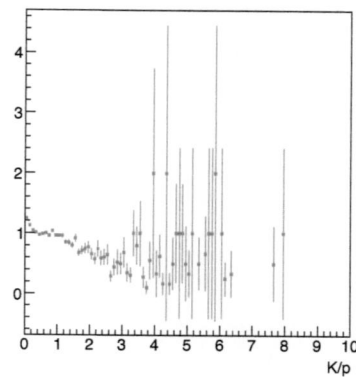

B.2. 10% BIN WIDTH

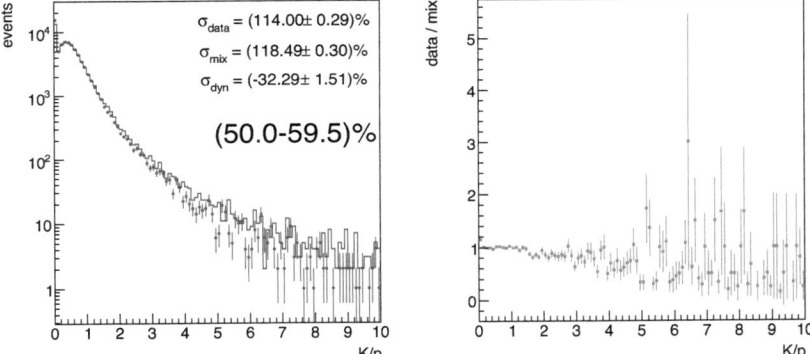

Figure B.6: Distributions of the event-by-event K/p ratio for data and mixed events for 10% centrality bins of minimum bias Pb + Pb collisions at 158A GeV. Right panels show the ratio of data to mixed events.

Bibliography

[1] A. Andronic, P. Braun-Munzinger and J. Stachel, Nucl. Phys. A 772 (2006) 167 [arXiv:nucl-th/0511071].

[2] Quark Matter 2004, J. Phys. G: Nucl. Part. Phys. (2004) 30.

[3] Z. Fodor and S. D. Katz, JHEP 0404 (2004) 050 [arXiv:hep-lat/0402006].

[4] F. Weber, J. Phys. G: Nucl. Part. Phys. 27 (2001) 465.

[5] F. Wilczek, Physics Today 53 (2000) 22 and hep-ph/0003183.

[6] F. Karsch et al., Nucl. Phys. B 502, (2001) 321.

[7] H. Stöcker and W. Greiner, Phys. Rep. 137 (1986) 277.

[8] V.D. Toneev et al., nucl-th/0309008.

[9] Z. Fodor and S.D. Katz, hep-lat/0402006, and JHEP 0404 (2004) 050.

[10] S. Ejiri et al. hep-lat/0312006.

[11] K. Suzuki et al., Phys. Rev. Lett. 92 (2004) 072302.

[12] Y. Shin et al., Phys. Rev. Lett. 81 (1998) 1576.

[13] F. Laue, C. Sturm et al., Phys. Rev. Lett. 82 (1999) 1640.

[14] P. Crochet et al., Phys. Lett. B 486 (2000) 6.

[15] R.J. Porter, Phys. Rev. Lett. 79 (1997) 1229.

[16] G. Agakichiev, Phys. Lett. B 422 (1998) 405.

[17] F. Klingl et al., Phys. Rev. Lett. 82 (1999) 3396.

[18] Cern Courier, April 2003, p.13.

[19] M. Gyulassi, J. Phys. G: Nucl. Part. Phys. 30 (2004) S911.

[20] M. Stephanov, K. Rajagopal, E. Shuryak, Phys. Rev. Lett. 81 (1998) 4816

[21] R. Rapp and J. Wambach, Nucl. Phys. A 661 (1999) 33c.

[22] T. Matsui and H. Satz, Phys. Lett. B 178 (1986) 416.

[23] M. Abreu et al., Phys. Lett. B 477 (2000) 28.

[24] J. Rafelski and B. Müller, Phys. Rev. Lett. 48 (1982) 1066

[25] G.E. Bruno et al., J. Phys. G: Nucl. Part. Phys. 30 (2004) 717c.

[26] M. Gazdzicki et al., J. Phys. G: Nucl. Part. Phys. 30 (2004) 701c.

[27] C.R. Allton et al., Phys. Rev. D 68, 014507 (2003)

[28] C. Alt et al. [NA49 Collaboration], Phys. Rev. C 78 (2008) 034914 [arXiv:0712.3216 [nucl-ex]].

[29] B. Lungwitz et al. [NA49 Collaboration], PoS CPOD07 (2007) 023 [arXiv:0709.1646 [nucl-ex]].

[30] T. Anticic et al. [NA49 Collaboration], Phys. Rev. C 79 (2009) 044904 [arXiv:0810.5580 [nucl-ex]].

[31] K. Grebieszkow et al., PoS CPOD07 (2007) 022 [arXiv:0707.4608 [nucl-ex]].

[32] C. Alt et al., arXiv: 0808.1237v3 [nucl-ex] (2009)

[33] B. I. Abelev et al. [STAR Collaboration], arXiv:0901.1795 [nucl-ex].

[34] B.I. Abelev et al., arXiv: 0901.1795v1 [nucl-ex] (2009)

[35] S. A. Bass et al. Prog. Part. Nucl. Phys. 41 (1998) 225-370.

[36] M. Bleicher et al. J. Phys. G: Nucl. Part. Phys. 25 (1999) 1859-1896.

[37] http://th.physik.uni-frankfurt.de/ urqmd/.

[38] S. Afanasiev et al. Nucl. Instr. Meth. A 430 (1999) 210-244.

[39] J. N. Marx and D. R. Nygren, Phys. Today 31N10 (1978) 46.

[40] B. Lasiuk, Nucl. Instrum. Meth. A 409 (1998) 402.

[41] S. V. Afanasiev et al., Phys. Rev. C 66 (2002) 054902.

[42] C. Roland, PhD thesis, J. W. G. Universität, Frankfurt am Main (1999).

[43] T. Schuster, PhD thesis, J. W. G. Universität, Frankfurt am Main, to be submitted in 2010.

[44] A. Laszlo, NA49 note (2006)

[45] C. Roland J. Phys. G: Nucl. Part. Phys. 31 (2005) 1075-1078.

[46] S. Jeon, V. Koch Phys. Rev. Lett. 83 (1999) 5435-5438.

[47] CBM Technical Status Report (2007).

[48] M. Deveaux et al., arXiv:0906.1301 [nucl-ex].

[49] J. M. Heuser, Nucl. Instrum. Meth. A 582 (2007) 910.

[50] J. M. Heuser, M. Deveaux, C. Muntz and J. Stroth, Nucl. Instrum. Meth. A 568 (2006) 258.

[51] J. M. Heuser, W. F. J. Muller, P. Senger, C. Muntz and J. Stroth, AIP Conf. Proc. 842 (2006) 1073.

[52] C. Hohne et al., Nucl. Instrum. Meth. A 595 (2008) 187.

[53] A. Andronic, Nucl. Instrum. Meth. A 563 (2006) 349.

[54] A. Akindinov et al., Nucl. Instrum. Meth. A 572 (2007) 676.

[55] I. E. Korolko and M. S. Prokudin, Phys. Atom. Nucl. 72 (2009) 293.

[56] A. Kiseleva, P. Senger and I. Vassiliev, Phys. Part. Nucl. 39 (2008) 1090.

[57] I. Kisel, Nucl. Instrum. Meth. A 566 (2006) 85.

Die VDM Verlagsservicegesellschaft sucht für wissenschaftliche Verlage abgeschlossene und herausragende

Dissertationen, Habilitationen, Diplomarbeiten, Master Theses, Magisterarbeiten usw.

für die kostenlose Publikation als Fachbuch.

Sie verfügen über eine Arbeit, die hohen inhaltlichen und formalen Ansprüchen genügt, und haben Interesse an einer honorarvergüteten Publikation?

Dann senden Sie bitte erste Informationen über sich und Ihre Arbeit per Email an *info@vdm-vsg.de*.

Sie erhalten kurzfristig unser Feedback!

VDM Verlagsservicegesellschaft mbH
Dudweiler Landstr. 99
D - 66123 Saarbrücken

Telefon +49 681 3720 174
Fax +49 681 3720 1749

www.vdm-vsg.de

Die VDM Verlagsservicegesellschaft mbH vertritt

Printed by Books on Demand GmbH, Norderstedt / Germany